AQA GCSE (9–1)
Biology

Achieve Grade 8–9 Workbook

Heidi Foxford
Shaista Shirazi

William Collins' dream of knowledge for all began with the publication of his first book in 1819. A self-educated mill worker, he not only enriched millions of lives, but also founded a flourishing publishing house. Today, staying true to this spirit, Collins books are packed with inspiration, innovation and practical expertise. They place you at the centre of a world of possibility and give you exactly what you need to explore it.

Collins. Freedom to teach

HarperCollins Publishers
The News Building
1 London Bridge Street
London SE1 9GF

**Browse the complete Collins catalogue at
www.collins.co.uk**

First edition 2016

10 9 8 7 6 5

© HarperCollins Publishers 2016

ISBN 978-0-00-819433-8

Collins® is a registered trademark of HarperCollins Publishers Limited

www.collins.co.uk

A catalogue record for this book is available from the British Library

Written by Heidi Foxford, Shaista Shirazi and Mike Smith
Commissioned by Joanna Ramsay
Project managed by Sarah Thomas and Siobhan Brown
Copy edited by Rebecca Ramsden
Proofread by Helen Bleck
Answer check by Claire-Coombe Jones
Typeset by Jouve India Pvt Ltd.,
Artwork by Jouve India Pvt Ltd.
Cover design by We are Laura and Jouve
Cover image: Spyros/Shutterstock, Dudarev Mikhail/Shutterstock
Printed by Grafica Veneta S.p.A

Contents

Section 1 ● Cell biology

Plant and animal cells (eukaryotic cells)   4
Bacterial cells (prokaryotic cells)   5
Size of cells and cell parts   6
Electron microscopes   7
Growing microorganisms   8
Cell specialisation and differentiation   9
Cell division by mitosis   10
Stem cells   12
Diffusion in and out of cells   13
Exchange surfaces in animals   14
Osmosis   16
Active transport   18

Section 2 ● Organisation

Digestive system   20
Digestive enzymes   21
Factors affecting enzymes   22
The heart and blood vessels   23
Blood   24
Heart–lungs system   25
Coronary heart disease   26
Risk factors for non-infectious diseases   27
Cancer   29
Leaves as a plant organ   30
Transpiration   31
Translocation   32

Section 3 ● Infection and response

Microorganisms and disease   33
Viral diseases   34
Bacterial diseases   35
Malaria   36
Human defence systems   37
Vaccination   39
Antibiotics and painkillers   40
Making and testing new drugs   41
Monoclonal antibodies   42
Plant diseases   43
Identification of plant diseases   44
Plant defence responses   45

Section 4 ● Photosynthesis and respiration reactions

Photosynthesis reaction   46
Rate of photosynthesis   47
Limiting factors   49
The uses of glucose from photosynthesis   50
Cell respiration   51
Anaerobic respiration   52
Response to exercise   53

Section 5 ● Automatic control systems in the body

Homeostasis   54
The nervous system and reflexes   55
The brain   56
The eye   57
Control of body temperature   59
Hormones and the endocrine system   60
Controlling blood glucose   61
Maintaining water and nitrogen balance in the body   62
Hormones in human reproduction   63
Contraception   65
Using hormones to treat human infertility   66
Negative feedback   67
Plant hormones   68
Uses of plant hormones   69

Section 6 ● Inheritance, variation and evolution

Sexual and asexual reproduction   71
Cell division by meiosis   72
DNA, genes and the genome   73
Structure of DNA   75
Protein synthesis and mutations   76
Inherited characteristics   78
Inherited disorders   80
Sex chromosomes   81
Variation and mutations   82
Theory of evolution   83
Speciation   84
The understanding of genetics   86
Fossil evidence   87
Other evidence for evolution   88
Extinction   89
Selective breeding   90
Genetic engineering   92
Cloning   93
Classification   95

Section 7 ● Ecology

Habitats and ecosystems   96
Food in the ecosystem   97
Abiotic and biotic factors   98
Adapting for survival   100
Measuring population size and species distribution   101
Cycling materials   103
Decomposition   104
Changing the environment   106
Effects of human activities   107
Global warming   108
Maintaining biodiversity   109
Biomass in an ecosystem   110
Food security   112
Role of biotechnology   113

Answers

Introduction

This workbook will help you build your confidence in answering Biology questions for GCSE Biology and GCSE Combined Science.

It gives you practice in using key scientific words, writing longer answers, answering synoptic questions as well as applying knowledge and analysing information.

You will find all the different question types in the workbook so you can get plenty of practice in providing short and long answers.

---

5. Describe, in detail, how carbon is continuously cycled through the ecosystem.

_____
_____
_____
_____
_____
_____

[6 marks]

Decomposition

1. The instruction manual for a garden compost bin recommends putting the bin in a sunny area and keeping the lid on. It also suggests turning the compost every month.

Explain why this advice is given.

_____
_____
_____
_____

[3 marks]

2. A class investigates the effect of temperature on the rate of decay of fresh milk by measuring pH change. Their method is shown below:

Required practical

1. Add 6 drops of phenolphthalein to a test tube; then add 5 cm³ of milk and 7 cm³ of sodium carbonate solution.
2. Place test tube in water bath until contents reach same temperature.
3. Add 1 cm³ of lipase into the test tube and start the stop clock.
4. Stir the contents of the test tube until the solution loses its pink colour and record the time.
5. Repeat steps 1–4 for a range of different temperatures.

stirring rod

1 ml lipase solution

add in turn:
5 ml milk
7 ml sodium carbonate solution
5 drops of phenolphthalein

stir and start timing when you add the lipase

104

The questions also cover required practicals, maths skills and synoptic questions – look out for the tags which will help you to identify these questions.

Higher Tier content is clearly marked throughout.

3. How can monoclonal antibodies be used in the treatment of cancer?

Higher Tier only

_____
_____

[3 marks]

Plant diseases

1. A gardener notices purple and black spots on the leaves of her rose plants. Some leaves have turned yellow and dropped off.

a Suggest the disease that may be affecting the rose plant.

_____ [1 mark]

b How does this disease spread?

_____ [1 mark]

c What is the treatment for this disease?

_____
_____ [2 marks]

2. Tobacco mosaic virus (TMV) causes leaf discolouration.

Worked example

Explain why TMV causes plants to have stunted growth. [4 marks]

TMV causes a lack of nutrition and a low immune system in plants. The virus reproduces in the cells causing growth issues.

Marks gained: [0 marks]

This answer would not gain any marks. The student should have linked the discolouration of the leaves (in the question) with a lack of chlorophyll/ability to photosynthesise. Less photosynthesis means less glucose made; therefore less amino acids/proteins/cellulose for growth; because glucose is needed to make amino acids/proteins/cellulose.

3. Aphids are an example of insect pests that affect plants.

Explain how aphids reduce crop yield.

_____
_____ [2 marks]

43

Learn how to answer test questions with annotated worked examples.

This will help you develop the skills you need to answer questions.

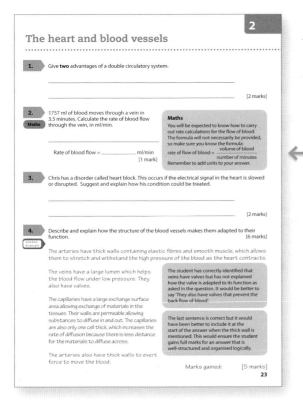

The amount of support gradually decreases throughout the workbook. As you build your skills you should be able to complete more of the questions yourself.

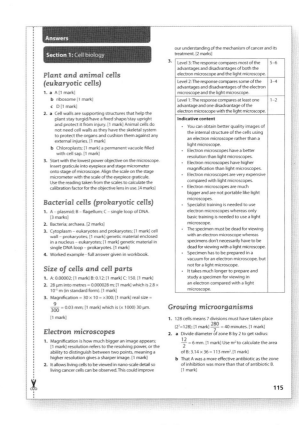

There are answers to all the questions at the back of the book. You can check your answers yourself or your teacher might tear them out and give them to you later to mark your work.

Plant and animal cells (eukaryotic cells)

1. The diagram shows a cell.

a Which letter shows the structure responsible for controlling the amount of oxygen that enters and leaves the cell?

_____   [1 mark]

b Name the organelles where proteins are synthesised.

_____   [1 mark]

c Which letter shows the organelle that generates energy for the cell's activities?

_____   [1 mark]

2. **a** Suggest a reason why plant cells have a cell wall, but animal cells do not.

_____

_____   [2 marks]

b In addition to the cell wall, what other organelles would you see in a palisade cell taken from a sunflower leaf, that you wouldn't see in an animal cell?

_____

_____   [2 marks]

3. Matilda is using a light microscope to view cheek cells. She uses an eyepiece graticule.

Required practical

Describe how Matilda could use an eyepiece graticule to measure the size of a cheek cell.

Maths

As most cells are very small, their sizes are written in standard form. Make sure you are confident using standard form.

_____

_____

_____

_____   [4 marks]

Bacterial cells (prokaryotic cells)

1. The diagram shows a prokaryotic cell.

Name structures A, B and C shown in the diagram.

A: _____

B: _____

C: _____ [3 marks]

2. Give **two** examples of prokaryotes.

_____ [2 marks]

3. Complete the table. For each feature, tick whether it is found in eukaryotic cells, prokaryotic cells or both. The first one has been done for you. [4 marks]

Feature	Eukaryotes	Prokaryotes
Cell membrane	✓	✓
Cytoplasm		
Cell wall		
Genetic material enclosed in a nucleus		
Genetic material in single DNA loop		

4. Carl Woese, an American microbiologist, proposed the 'Three-domain classification system'. This places archaea in a separate domain to eukaryotes and bacteria.

Worked Example

Evaluate the evidence for the three-domain system. [3 marks]

Chemical analysis of archaea has shown that they have DNA very different bacteria. The nucleic acid in the archaea was found to be more closely related to eukaryotes than to prokaryotes. As a result, Carl Woese suggested they should be in a completely separate domain. The evidence is very strong as it is based on chemical analysis which has been replicated by other scientists. Other scientists have not found evidence against Woese's theory.

'Evaluation' questions require a response that considers the evidence for and against; this answer would gain full marks.

Marks gained: [3 marks]

Size of cells and cell parts

1. The table shows sizes of different plant and animal cells and their organelles.

Maths

Cell	Size	Size in mm
Leaf cell	70 µm	0.07
Plant cell ribosome	20 nm	A
Human egg cell	120 µm	B
Egg cell mitochondria	2 µm	0.002
Ostrich egg cell	15 cm	C

Calculate the size of A, B and C. _____

A : _____ mm

B : _____ mm

C : _____ mm   [3 marks]

2. A skin cell is 28 µm in diameter.

Maths

Write this in standard form. Give your answer in metres. Show your workings.   [2 marks]

_____

= _____ m

3. John uses an eyepiece lens with a magnification of ×10 and an objective lens with a magnification of ×30. The image size of the specimen is 9 mm.

Maths

Calculate the real size of the specimen. Write your answer in micrometres.   [3 marks]

_____

_____

= _____ µm

Maths

Remember, if you are converting a smaller unit to a larger unit, your number should get smaller; and if you are converting a larger unit into a smaller unit, your number should get bigger.

Electron microscopes

1. Explain the difference between magnification and resolution.

_____

_____

_____ [2 marks]

2. In 2014, a group of scientists won the Nobel Prize for development of super-resolved fluorescence microscopy. It allows a much higher resolution than light microscopy and allows living cells to be viewed. Suggest how the development of the super-resolved fluorescence microscope could improve our understanding of cancer.

_____

_____

_____ [2 marks]

3. Mitochondria can be viewed using ether an electron microscope or a light microscope.

Compare the advantages and disadvantages of using an electron microscope rather than a light microscope.

_____

_____

_____

_____

_____

_____ [6 marks]

Growing microorganisms

1.

Maths

One *E. coli* cell produces 128 cells in 280 minutes.

What is the mean division time of *E. coli*?

_____

Mean division time = _____ minutes   [2 marks]

2.

Required practical

Dottie is investigating the ability of two antibiotics to kill *E. coli* bacteria. The diagram below shows her results after 24 hours of incubation.

The zone of inhibition for antibiotic A was 314 mm².

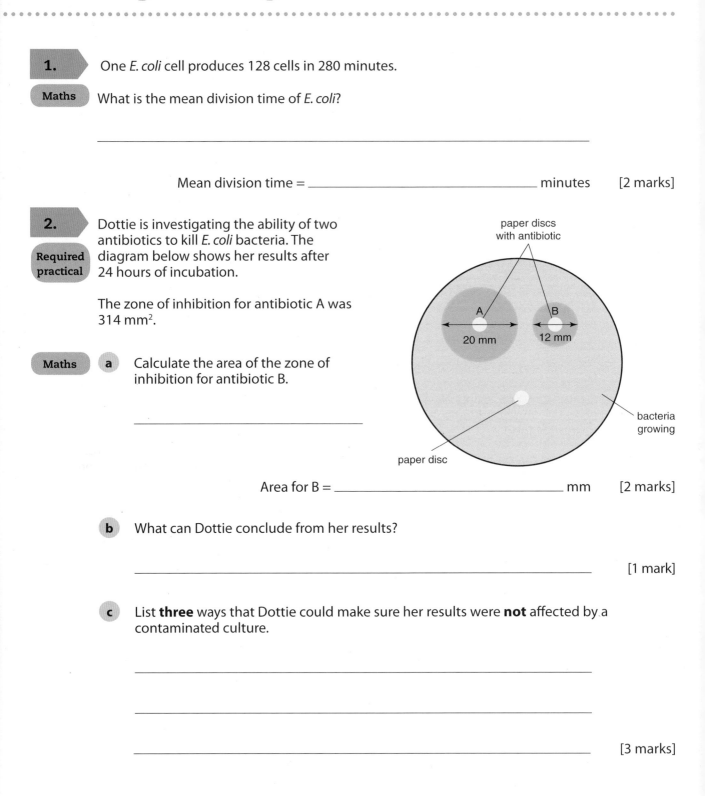

paper discs with antibiotic

bacteria growing

paper disc

Maths

a   Calculate the area of the zone of inhibition for antibiotic B.

_____

Area for B = _____ mm   [2 marks]

b   What can Dottie conclude from her results?

_____   [1 mark]

c   List **three** ways that Dottie could make sure her results were **not** affected by a contaminated culture.

_____

_____

_____   [3 marks]

Cell specialisation and differentiation

1. Explain the importance of cell differentiation.

_____

_____ [2 marks]

2. Explain how **both** human gametes are adapted to ensure successful growth and development of an embryo.

_____

_____

_____ [2 marks]

3. Acinar cells are found in the pancreas.

They produce and transport enzymes that are passed into the duodenum, where they assist in the digestion of food.

Suggest and explain how many ribosomes you would expect to find in an acinar cell compared with other cells.

Command words

The command word 'suggest' requires you to **apply** your knowledge and understanding to a new situation, so you are not expected to have learnt the answers to these questions. Use your knowledge and understanding of cell structure and function to make a sensible suggestion.

_____

_____

_____ [3 marks]

4. Compare cell differentiation in plant cells to cell differentiation in animal cells.

_____

_____

_____ [2 marks]

Cell division by mitosis

1. Explain the difference between a gene and a chromosome.

_____

_____ [2 marks]

2. The diagram shows the cell cycle.

The descriptions in the table below describe the stages in the cell cycle. Complete the table to show which numbered stage of the cell cycle in the diagram matches each description.

The descriptions in the table are **not** in the correct order.

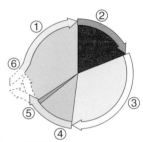

Description of stage	Number on cell cycle diagram
The cell grows and the number of sub-cellular structures increases	1
The cytoplasm divides into two and the new cell membrane forms two new cells	
Further growth occurs and the DNA is checked for errors	
Mitosis occurs and the chromosomes move apart, forming two new nuclei	
The DNA replicates	
Temporary resting period or the cell no longer divides	6

[4 marks]

3. In a sample of 170 cells, 14 are observed to be in the last stage of mitosis.

Maths

A complete cell cycle takes 23 hours.

How long does the last stage of mitosis take for an average cell in this tissue? Give your answer in hours, rounded to two decimal places.

Maths

The fraction of cells in a particular stage of the cell cycle is proportional to the time taken for that stage. Therefore, if you know the **number of cells in a particular stage**, the **total number of cells** and the **total time for the cell cycle**, you can work out an estimate for how long that particular stage takes.

_____

_____

Answer = _____ hours   [3 marks]

4.

Scientists calculate the mitotic index to identify cancerous cells. The mitotic index is the ratio of cells in a tissue that are undergoing mitosis and it can be used as an indication of the rate of growth.

Table A

Type of cell	Average percentage of cells at rest	Mitotic index
Lung	95	5
Stomach	90	10
Ovary	85	15

Table B

Type of cell	Average percentage of cells at rest	Mitotic index
Lung	80	20
Stomach	70	30
Ovary	55	45

a Identify which table shows cancerous cells. Use evidence from the tables to explain your answer. [3 marks]

Table B shows cancerous cells. Because cancer cells have a higher mitotic index because more of the cells are reproducing out of control. This means scientists would see a lower percentage of cells being at rest, which is a characteristic of cell division in cancerous cells.

Marks gained: [2 marks]

The student has correctly identified that Table B shows cancerous cells. The explanation is correct, but the student has not used evidence from the table, as asked, and would therefore lose marks.

b Use the data shown to determine which type of cell is dividing most rapidly. [1 mark]

Ovary cells

Marks gained: [1 mark]

Correct. Ovary cells have the lowest percentage of cells at rest and therefore the highest mitotic index, caused by the cells reproducing so fast.

Maths **c** Use the data in the tables to determine the mitotic index if the percentage of cells at rest is 45%. [1 mark]

Mitotic index = ___30___

Marks gained: [0 marks]

Incorrect. The question asks you to 'use the data in the tables to determine the mitotic index', so you need to look for any patterns in the data. For every 10% reduction in the cells at rest, there is a 10-point increase in the mitotic index; therefore, if the percentage of cells at rest was 45%, the mitotic index would be 55.

Stem cells

1. What are stem cells and why are they important?

_____

_____ [2 marks]

2. Which of the following statements about stem cells are true? Tick **two** boxes.

☐ Stem cells are only found in embryos.

☐ Stem cells can be grown in a laboratory to form clones.

☐ One stem cell can produce more stem cells.

☐ Stem cells never mutate. [2 marks]

3. State **two** benefits of using stem cells from meristems in plants to produce clones of plants.

_____

_____ [2 marks]

4. Dutch elm disease is a fungal tree disease that has destroyed millions of elm trees in the UK. Suggest how meristem cells could play a role in preventing extinction of Dutch elm trees.

Synoptic

_____

_____ [2 marks]

5. Evaluate the advantages and disadvantages associated with therapeutic cloning using embryonic stem cells.

_____

_____

_____

_____

_____

_____ [6 marks]

Diffusion in and out of cells

1. Define the term diffusion.

_____ [1 mark]

2. Explain why the rate of diffusion of carbon dioxide into stomata on a leaf is higher on a warm day.

_____

_____ [2 marks]

3. Describe and explain the movement of oxygen across the cell membrane of an active muscle cell during aerobic respiration.

_____

_____

_____ [3 marks]

4. The table shows three model cells with different concentrations of glucose inside and outside each cell. Each cell is represented by a dashed circular line.

Complete the table. For each cell, **A**, **B** and **C**:

a draw an arrow to show the direction of net movement of glucose molecules;

provide an explanation for the arrow you have drawn.

Cell	Direction of movement	Explanation
A	1.0 mol dm³ 0.5 mol dm⁻³	
B	1.0 mol dm³ 1.0 mol dm⁻³	
C	0.5 mol dm³ 2.0 mol dm⁻³	

[6 marks]

b In which cell will the rate of diffusion be fastest? Explain your answer.

_____

_____

[2 marks]

Remember

As diffusion takes place, the concentration gradient between the two sides of a partially permeable membrane decreases so the rate of diffusion will slow down until both sides become equal. At this point the particles will still be moving but there will be **no net** (overall) movement between the sides.

5.

Emphysema is a long-term, progressive disease of the lungs that causes shortness of breath, enlargement of the alveoli, thicker alveolar walls and loss of elasticity in the lungs.

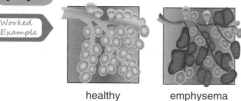

healthy   emphysema

Suggest and explain why emphysema causes shortness of breath.   [5 marks]

Enlarged alveoli result in decreased surface area of alveoli, which reduces available space for diffusion. This means the rate of diffusion of oxygen crossing into capillaries surrounding alveoli is limited.

Also, thickening of alveolar walls increases the distance oxygen has to diffuse, which would also reduce the rate of diffusion of oxygen into the blood.

Both of these things mean concentration of oxygen in the blood is less than normal so less oxygen is available to cells for respiration and the person gets short of breath.

Marks gained: [4 marks]

This is a good answer with excellent use of scientific words and a clear explanation.

Notice that the information given in the question also mentions loss of elasticity of lungs, but this answer has not referred to how this feature of emphysema could contribute to shortness of breath, missing out on a mark. The answer needs to include a point about reduced elasticity causing old air to get trapped in alveoli. This reduces the concentration gradient for the diffusion of oxygen from the alveoli into the capillaries in the lungs and is another reason why the rate of diffusion of oxygen is lower than normal.

Exchange surfaces in animals

1.

Which of the following organisms would have the **lowest** surface area to volume ratio?

Tick **one** box.

	Slug		Mouse
	Camel		Hippopotamus [1 mark]

Maths

You need to know how to calculate the surface area to volume ratio. Make sure you are confident calculating surface area and volume. You should also know the formula:

$$\text{Surface area to volume ratio} = \frac{\text{surface area}}{\text{volume}}$$

2. List **three** substances that move in and out of cells in the human body.

_____ [3 marks]

3. The surface area of a cell affects the rate at which particles enter and leave the cell.

Maths The table shows different-sized cubes that represent cells.

a Complete the table by calculating the surface area, volume, and surface area to volume ratio (SA:V) for the $3 \times 3 \times 3$ cm cube.

Letter of organism	Dimensions of organism (cm)	Surface area (cm²)	Volume (cm³)	SA:V
A	⌷ } 1 cm	6	1	6:1
B	} 2 cm	24	8	3:1
C	} 3 cm	_____	_____	_____
D	} 4 cm	96	64	1.5:1

[3 marks]

b Explain which organism (A, B, C or D) is likely to be most efficient at absorbing substances by diffusion.

_____ [1 mark]

4. Manatees have a very low metabolism, which makes them unable to heat their bodies well, so they have to stay in warm water.

Suggest how its round shape helps the manatee survive.

_____

_____ [2 marks]

5. Describe and explain how a mackerel's gills are adapted for gas exchange. [6 marks]

Worked Example

The gills are made of lots of feathery projections, which give a large surface area for the exchange of gases and increase the rate of diffusion.

Each projection is covered in tiny structures called lamellae, which increases the surface area even more. The lamellae have lots of blood capillaries to speed up diffusion between water and blood. They also have thin walls to reduce the distance the gases have to diffuse which increases the rate of diffusion.

> Try to use the correct scientific names for parts of the fish. The student is referring to the gill filaments, so it would have been better to use this term rather than 'feathery projections'.

The water flows through the lamellae in one direction and blood in the other direction and this keeps a large concentration gradient. As a result, the concentration of oxygen in the water is always higher than in the blood so the rate of diffusion of oxygen from the water into the blood is maximised.

> This answer is well-structured and would gain full marks for presenting ideas clearly and in a logical order. The student has provided clear explanations of the adaptations and uses scientific terms such as 'concentration gradient', demonstrating a good understanding of the science. This answer would gain full marks.

Marks gained: [6 marks]

Osmosis

1. Define osmosis.

_____

_____ [2 marks]

2. Tim investigated the effect of different concentrations of sugar solution on the mass of potato tissue.

Required practical

He cut 8 equal-sized potato chips. Then he measured and recorded the mass of each one.

Next he placed each chip into a boiling tube containing a different concentration of sugar.

He left them for 45 minutes before removing each one and weighing its mass again.

Tim calculated the change in mass and used this to calculate the percentage change in mass of each cylinder.

a Explain why Tim calculated the change in mass and then the percentage change in mass.

_____ [1 mark]

b Name **two** variables to be controlled in Tim's investigation.

_____ [2 marks]

Tim's results are shown below.

Boiling tube number	1	2	3	4	5	6	7	8
Concentration of sugar solution (M)	0.1	0.2	0.3	0.4	0.5	0.6	0.7	0.8
% change in mass of potato cylinder	6.5	5.0	3.0	1.0	−1.5	−3.5	−6.5	−9.0

Maths **c** Draw a graph to show concentration of sugar solution against percentage change in mass. Add a line of best fit.

[4 marks]

Maths **d** Use your graph to estimate the concentration of the solution inside the potato cells.

Concentration inside cells = _____ M   [1 mark]

e Explain the difference in the results between boiling tube 2 and boiling tube 7.

_____

_____

_____

_____

_____

[4 marks]

Active transport

1. Compare the processes of active transport and diffusion. [3 marks]

Worked Example

Active transport and diffusion have quite a few differences and some similarities.

Active transport happens against a concentration gradient whereas diffusion happens down a concentration gradient.

Active transport requires energy from respiration but diffusion doesn't. Active transport uses carrier molecules that are specific. Both can happen across a cell membrane but diffusion also happens in air and other liquids. Both are methods of transport of substances in living things.

Marks gained: [3 marks]

It is good practice to use 'lead-in' sentences when writing in science, but in an exam it is unnecessary. You will not gain any marks for an introduction to an answer or repeating the question, so try to use your time efficiently and get straight to the point.

This answer is clear, concise and well-structured and would gain full marks. It covers the differences and similarities in detail. It contains more differences and similarities than needed to pick up the 3 marks so the student could have saved time by not writing as much.

2. Explain why plants are **not** able to rely on diffusion or osmosis for the uptake of mineralions from the soil.

_____

_____

_____ [3 marks]

3. Root hair cells contain many mitochondria.

Explain how this feature of root hair cells might prevent plant mineral deficiencies.

_____

_____

_____ [3 marks]

4. Explain why we would starve if active transport did **not** take place in the gut.

_____

_____

Command words

If you are asked to 'explain' something, you must make clear the reasons why something might happen rather than just describing it.

_____

_____ [3 marks]

Digestive system

1. The digestive system is an organ system. Describe what is meant by an 'organ system'.

_____ [1 mark]

2. What are the products of digestion used for?

_____

_____ [2 marks]

3. Compare the function of the small intestine with that of the large intestine.

_____

_____ [2 marks]

4. The total length of the average digestive tract is 9 m.

Maths What is this in μm? Give your answer in standard form.

_____

= _____ μm [1 mark]

5. The gall bladder stores bile.

Describe and explain the function of bile.

_____

_____

_____ [3 marks]

6. Explain how villi increase the absorption of nutrients in the small intestine.

_____

_____

_____ [3 marks]

Digestive enzymes

1. Complete the table to give the actions of the different digestive enzymes.

Name of enzyme	Action
Carbohydrases	Break down carbohydrates into glucose
Proteases	
Lipases	

[2 marks]

2. **Required practical** A scientist tests a cake that is claimed to contain 'no fat'. She prepares a sample of the cake by grinding it up, adding distilled water, mixing the solution and then filtering it. Describe the method the scientist should use to find out if the cake contains fat.

_____

_____

_____

[3 marks]

3. **Required practical** Ted tests unknown solutions for reducing sugars using Benedict's solution. His results are shown in the table.

	Solution A	Solution B	Solution C	Solution D
Colour change observed	brick-red	green	yellow	blue

Put the solutions in order of sugar concentration, from lowest to highest. Use the letters **A**, **B**, **C** and **D** in your answer.

(lowest sugar concentration) _____ (highest sugar concentration)   [1 mark]

4. Sucrose is a type of sugar used in sweeteners. It produces a negative result when tested with Benedict's solution. Suggest why.

_____

_____

Practical

In the exam you could be asked how to carry out tests to detect the presence of sugars, starch, protein or lipids, and the colour changes to expect for a positive test. Think of ways to remember the colour changes, e.g. the Biuret test for **p**rotein is **p**ink or **p**urple if **p**ositive.

_____

_____

[2 marks]

Factors affecting enzymes

1. The lock and key theory is a model to explain how enzymes work.

Describe and explain the lock and key theory.

_____

_____

_____ [3 marks]

2. Sarah investigates the effect of temperature on the activity of the enzyme amylase. She uses a continuous sampling technique to determine the time taken to completely digest a starch solution at a range of temperatures. Every 30 seconds she uses iodine solution to test for starch.

Required practical

a Sarah uses a beaker of water and a Bunsen burner to heat the amylase.

Suggest why this could be a source of error in the results.

_____ [1 mark]

b Suggest how Sarah could improve this part of her method.

_____ [1 mark]

c Sarah's results are shown in the graph.

Suggest an explanation for each labelled part of the graph.

_____

_____

_____

_____

_____

_____ [6 marks]

The heart and blood vessels

1. Give **two** advantages of a double circulatory system.

_____

_____ [2 marks]

2. **Maths**

1757 ml of blood moves through a vein in 3.5 minutes. Calculate the rate of blood flow through the vein, in ml/min.

_____

Rate of blood flow = _____ ml/min

[1 mark]

Maths

You will be expected to know how to carry out rate calculations for the flow of blood. The formula will not necessarily be provided, so make sure you know the formula:

$$\text{rate of flow of blood} = \frac{\text{volume of blood}}{\text{number of minutes}}$$

Remember to add units to your answer.

3. Chris has a disorder called heart block. This occurs if the electrical signal in the heart is slowed or disrupted. Suggest and explain how his condition could be treated.

_____

_____ [2 marks]

4. Describe and explain how the structure of the blood vessels makes them adapted to their function. [6 marks]

Worked Example

The arteries have thick walls containing elastic fibres and smooth muscle, which allows them to stretch and withstand the high pressure of the blood as the heart contracts.

The veins have a large lumen which helps the blood flow under low pressure. They also have valves.

The capillaries have a large exchange surface area allowing exchange of materials in the tissues. Their walls are permeable allowing substances to diffuse in and out. The capillaries are also only one cell thick, which increases the rate of diffusion because there is less distance for the materials to diffuse across.

The arteries also have thick walls to exert force to move the blood.

The student has correctly identified that veins have valves but has not explained how the valve is adapted to its function as asked in the question. It would be better to say 'They also have valves that prevent the back flow of blood'.

The last sentence is correct but it would have been better to include it at the start of the answer when the thick wall is mentioned. This would ensure the student gains full marks for an answer that is well-structured and organised logically.

Marks gained: [5 marks]

Blood

. . . .

1. What is the function of plasma?

_____ [1 mark]

2. **Maths**

Approximately 55% of the blood is plasma. If a person has 6796 cm³ of blood in their body, how much would be plasma?

_____

Volume of plasma = _____ cm³   [1 mark]

3. Thrombocytopenia is a condition in which the blood has a lower than normal number of platelets.

a Describe the function of platelets.

_____

_____ [2 marks]

b Which of the following could be symptoms of thrombocytopenia? Tick **two** boxes.

☐ Excessive bruising   ☐ Irregular heart rate

☐ Excessive bleeding   ☐ Excessive bone fractures   [2 marks]

4. Describe and explain how red blood cells are adapted to transport oxygen.

_____

_____

_____

_____ [5 marks]

Heart–lungs system

1. Arteries usually carry oxygenated blood, but there is **one** exception. Name the artery that carries deoxygenated blood and explain why the blood it carries is deoxygenated.

_____

_____

_____ [3 marks]

2. Fabio takes 108 breaths in 12 minutes. Idris takes 91 breaths in 7 minutes. Calculate who has the faster breathing rate. Show your working.

Maths

_____

Person with fastest breathing rate: _____ [2 marks]

3. The diaphragm receives electrical impulses from the phrenic nerve.

Evie has damaged her phrenic nerve. Explain why this could be life-threatening.

_____

_____

_____ [3 marks]

4. Atrial septal defect (ASD) is a condition where there is a hole in the wall that divides the atrial chambers of the heart.

Suggest and explain how a hole between the atria would affect the transport of oxygen around the body.

_____

_____

_____ [3 marks]

Coronary heart disease

1. Describe what a stent is. Explain how it lowers risk of a heart attack in people with coronary heart disease.

_____

_____ [2 marks]

2. Give **two** disadvantages of using stents.

_____

_____ [2 marks]

3. Explain how statins work.

_____

_____

_____ [3 marks]

4. Dr Patel has an 80-year-old patient, Betty, who has a weak and leaky heart valve in her aorta. She is quite frail and suffers from high blood pressure.

Evaluate the risks and benefits for Betty to undergo treatment to replace her weak heart valve with a mechanical valve.

> **Remember**
> You could be asked to evaluate any of the treatments for heart disease, including artificial hearts, and the use of stents and valves, so make sure you know the advantages and disadvantages for each type of treatment.

_____

_____

_____

_____ [4 marks]

5.

Worked Example

Artificial hearts come in different designs. Doctors compared the results for patients who received two **different designs** of artificial heart, A and B. They recorded information 3 years after the artificial hearts were implanted. Their results are shown in the table:

Maths

Design of artificial heart	**Information recorded 3 years after the artificial hearts were implanted**		
	Number of patients surviving with artificial heart	Number of patients surviving but who required repair or replacement of artificial heart	Number of patients who died
A Implanted in 131 patients	67	14	50
B Implanted in 62 patients	9	25	28

Which type of artificial heart was more successful? Use calculations to support your answer. [4 marks]

Design A was more successful. I decided this because 67/131 or 51% survived without needing replacement of artificial heart, compared to only 9/62 or 15% for design B.

Only 14/131 or 11% required repair or replacement of artificial heart in the 3 years, compared to 25/62 or 40% for design B. 50/131 or 38% of patients died with design A, which was less than the 28/62 or 45% of patients recorded for design B.

This answer has correctly used the data in the graph to calculate percentages of those surviving, surviving but need repair or replacement, and those that died. Calculating the percentages allows the student to directly compare the data of the types of artificial heart.

This is a clear and well-structured answer showing that the student has fully analysed and understood the data.

Marks gained: [4 marks]

Risk factors for non-infectious diseases

1.

Define what is meant by the term 'risk factor' for a disease.

_____ [1 mark]

2. Suggest **two** reasons why non-communicable disease are more common in deprived areas of the UK.

_____

_____ [2 marks]

3. A causal mechanism has been scientifically proven for some risk factors for non-communicable diseases, but not for others.

Why is it **difficult** to prove a non-communicable disease is caused by a particular factor?

Common misconception

Remember that, just because a graph shows a correlation, it does not mean one thing is causing the other. There might be other factors involved.

_____

_____

_____ [3 marks]

4.

Maths

The graph shows incidence of coronary heart disease in different categories of hip-to-waist measurements and body mass index (BMI) scores.

A student concluded 'the higher your quartile for hip-to-waist ratio and BMI, the higher the risk of coronary heart disease'. Evaluate her conclusion.

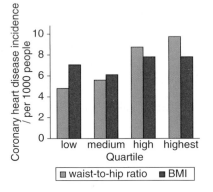

_____

_____

_____

_____ [3 marks]

Cancer

1. Compare benign and malignant tumours.

_____

_____

_____

_____ [4 marks]

2. Scientists investigated the relationship between fat intake in the diet and the death rate from breast cancer in 15 different countries. Their results are shown in the graph:

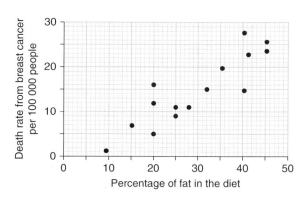

The graph is printed in a newspaper with the statement 'high percentage of fat in the diet causes breast cancer'.

Use evidence from the graph to evaluate this conclusion. [4 marks]

The graph shows a strong positive correlation. As the percentage of fat in the diet increases so does the death rate from breast cancer. But, correlation does not prove that the percentage of fat in the diet causes the death rate, it just suggests a link. There could be other factors involved.

However, the graph shows that, at some percentages of fat, e.g. 20% fat, there are three different death rates from three different countries. Also there are some countries with a higher death rate but which have a lower fat intake so these points go against the conclusion.

> The student has given evidence from the graph both for and against the conclusion printed in the paper. The student has correctly stated that, the positive correlation only suggests a link, it does not prove that increased fat intake causes breast cancer.

Marks gained: [4 marks]

Leaves as a plant organ

1. The diagram shows a plant leaf.

Name structures A, B and C.

A: _____

B: _____

C: _____

[3 marks]

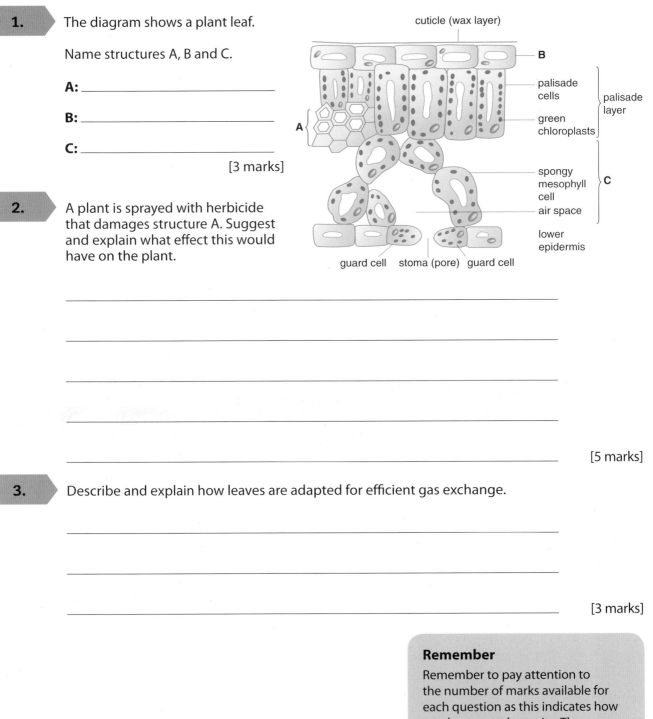

cuticle (wax layer)

B

palisade cells

palisade layer

A

green chloroplasts

spongy mesophyll cell

C

air space

lower epidermis

guard cell   stoma (pore)   guard cell

2. A plant is sprayed with herbicide that damages structure A. Suggest and explain what effect this would have on the plant.

_____

_____

_____

_____

_____

[5 marks]

3. Describe and explain how leaves are adapted for efficient gas exchange.

_____

_____

_____

[3 marks]

Remember

Remember to pay attention to the number of marks available for each question as this indicates how much you need to write. The space provided will also give you a sense of how long your answer should be.

Transipration

1. Explain why transpiration is important in plants.

_____

_____

_____

_____ [3 marks]

2. Erin investigates how temperature and humidity affect transpiration rates using a potometer, as shown in the diagram.

The table shows the volume of water produced in three different conditions.

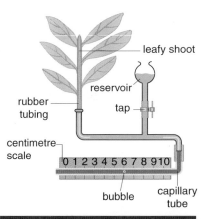

Conditions	Volume of water (ml)						Transpiration rate (ml/min)
	Time (mins)						
	0	5	10	15	20	25	
Normal conditions	0.0	0.4	1.0	1.4	1.9	2.3	
Increased air temperature	0.0	0.6	1.2	1.8	2.3	2.9	
Decreased humidity	0.0	0.8	1.3	2.0	2.7	3.4	

Maths **a** Complete the table by calcutating the transpiration rate for each condition. [1 mark]

b Explain the trends shown in the results for increased air temperature and decreased humidity.

_____

_____

_____

_____ [4 marks]

Translocation

1. Which statements about phloem are true? Tick **two** boxes.

☐ Phloem tubes are living cells joined end to end with small pores.

☐ Phloem transports food substances in both directions.

☐ Phloem transports food substances in only direction only.

☐ Phloem are only found in the leaves. [2 marks]

2. Marram grass grows on sand dunes. It has been found to have stomata located in sunken pits. Suggest and explain how these pits may help the plant survive. [2 marks]

Worked Example

Pits trap water vapour therefore the concentration gradient of water vapour between the outside and inside of the leaf is reduced so the rate of diffusion of water out of the stomata decreases. This reduces water loss through transpiration and therefore helps the plant survive.

This is a well-written answer that uses correct scientific terminology and contains the relevant points, showing the student has understood the question.

Marks gained: [2 marks]

3. Describe and explain the processes of transpiration and translocation.

_____

_____

_____

_____

_____

_____ [6 marks]

Literacy
For the 6-mark questions, you are assessed on your ability to structure your answer and present your ideas or explanations in a logical order. Although it might be easier to just write in bullet points, you are unlikely to get full marks as bullet points do not allow you to link ideas as writing in full sentences does.

Microorganisms and disease

1. **a** Explain what is meant by a communicable disease and give an example.

_____

_____

_____ [3 marks]

b State **one** way pathogens cause harm in the body.

_____ [1 mark]

2. Draw **one** line from each disease to the correct description.

Gonorrhoea	Attacks the body's immune system
	Can reduce spread by using handkerchiefs
Malaria	Caused by a fungus
	Caused by a protist
	Spread through unhygienic food preparation
Measles	Treated with antibiotics

[3 marks]

3. Compare the spread of bacteria and viruses in infectious diseases.

_____

_____

_____

_____

_____ [6 marks]

Literacy

This is an extended response question worth 6 marks, and requires you to link ideas to produce a coherent and structured comparison of the spread of infectious diseases by bacteria and viruses. Marks are awarded for levels of response. An error often repeated by students is that they make a number of points about the spread of disease by viruses and bacteria but fail to make a comparison between the two.

Viral diseases

1. Measles is caused by a virus.

a What are the symptoms of measles?

_____ [1 mark]

b How is measles spread?

_____ [1 mark]

c Explain why measles is a **serious** disease.

_____

_____ [2 marks]

2. Read the information.

> Measles can be prevented with the MMR vaccine.
>
> In 2000, measles caused 550,100 deaths worldwide. By 2016 this had decreased by 84%. This was due to an increase in the use of the MMR vaccine.
>
> In 2000, 72% of the world's children had received the MMR vaccine before their first birthday. In 2016, this had risen to 85%.

Maths
Remember, a percentage means the number out of 100.

Maths **a** Calculate the number of deaths due to measles in 2016.

_____

_____

Number of deaths due to measles in 2016 = _____ [2 marks]

Synoptic **b** The risk of catching measles decreases as the percentage vaccinated increases. This even applies to those who are **not** vaccinated. Suggest **one** reason why.

_____

_____ [1 mark]

3. **a** What is the name given to late stage HIV?

_____ [1 mark]

b Which cells of the body are attacked by the HIV?

_____ [1 mark]

c Name the type of drugs used to treat HIV.

_____ [1 mark]

Bacterial diseases

1. Jane thinks that bacteria which cause diseases in animals are all the same shape and size.

Is she correct? Explain why.

_____

_____ [2 marks]

2. **a** *Salmonella* food poisoning is spread by bacteria ingested in food.

Name **three** types of food that may contain *Salmonella*.

_____

_____ [2 marks]

b How is gonorrhoea spread?

_____ [1 mark]

c How can the spread of gonorrhoea be prevented?

_____

_____ [2 marks]

3. People with bacterial infections do **not** develop symptoms as soon as they are infected with the pathogen.

Explain the stages of infection with bacterial pathogens.

____

____

____

____

____ [4 marks]

Common misconception

There is a lag between the time at which an organism is infected and the symptoms of the disease appearing. This is because bacteria take time to acclimatise to the surroundings and reproduce rapidly. When there are significant numbers of bacteria, the toxins they produce will cause symptoms of infection to appear.

Malaria

1. The diagram shows stages in the life cycle of the malarial parasite.

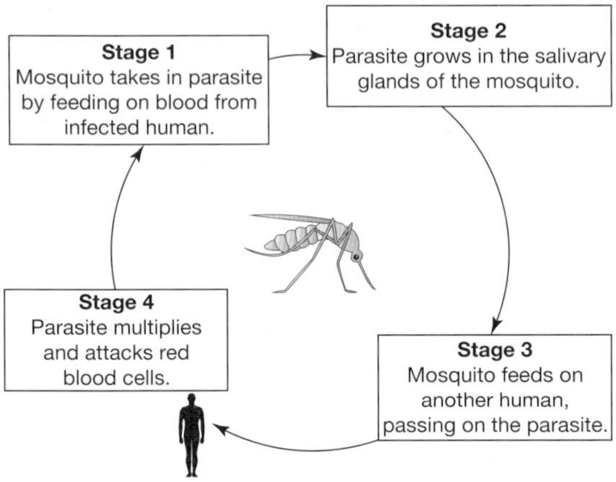

Stage 1
Mosquito takes in parasite by feeding on blood from infected human.
Stage 2
Parasite grows in the salivary glands of the mosquito.
Stage 3
Mosquito feeds on another human, passing on the parasite.
Stage 4
Parasite multiplies and attacks red blood cells.

a Which organism is a pathogen? Tick **one** box.

☐ Human ☐ Mosquito ☐ Parasite [1 mark]

b Which stages are prevented by the use of mosquito nets?

____ [1 mark]

c The parasites are transmitted by mosquitoes.

What name is given to organisms that **spread** disease rather than causing it? Tick **one** box.

☐ Insects ☐ Parasites ☐ Protists ☐ Vectors [1 mark]

2. Explain **two** ways that malaria can be controlled.

_____

_____

_____

_____ [4 marks]

Human defence systems

1. **a** How does stomach acid defend the body against pathogens?

_____ [1 mark]

b Name **two** ways the skin defends against pathogen attack.

_____

_____ [2 marks]

c Name **two** ways that the trachea defends against pathogen attack.

_____

_____ [2 marks]

2. Describe how platelets help prevent pathogens from entering a wound.

_____

_____

_____

_____ [4 marks]

3. Explain how bacteria make us feel ill once they enter the body.

_____

_____

_____ [3 marks]

4.

Synoptic Some white blood cells ingest and digest pathogens.

Describe what happens to any proteins the pathogens contain.

_____

_____ [2 marks]

5. Describe and explain how white blood cells help to defend against pathogens.

Common misconception

Antibodies produced in immune responses do **not** wait around for years to deal with invading microbes. The correct idea is that white cells gain the ability to produce antibodies quickly in response to invasion.

_____

_____

_____

_____

_____

_____

_____

_____

_____ [6 marks]

Vaccination

1. Explain why it can be more difficult to kill viruses inside the body than bacteria inside the body.

_____

_____ [2 marks]

2. **a** Describe what a vaccine contains.

_____

_____ [1 mark]

> **Common misconception**
> The suggestion that the 'disease' is injected is wrong. It is a form of the pathogen that causes the disease that is injected.

b To prevent epidemics of infectious diseases, a high percentage of the population should be vaccinated. Explain why a higher percentage of the population needs to be vaccinated against the most infectious diseases.

_____

_____ [2 marks]

3. Describe the immune response to vaccination.

_____

_____

_____ [3 marks]

4. The graph shows antibody level in blood after vaccination (**A**) and a second exposure to the same pathogen (**B**).

Explain why the increase in antibody level after **A** is different from the increase in antibody level after **B**.

_____

_____

_____ [3 marks]

Antibiotics and painkillers

1. Painkillers will **not** cure an infectious disease. Explain why.

_____

_____ [2 marks]

2. Scientists wanted to compare the effectiveness of two painkillers, drug **A** and drug **B**.

They chose 100 volunteers who were suffering pain.

They gave half the volunteers a dose of drug **A** and the other half a dose of drug **B**.

The volunteers recorded how much pain they felt in the next 24 hours.

a Is the test valid? Explain your answer.

_____

_____

_____ [2 marks]

b Suggest **two** factors that should be matched in the volunteers.

1 _____

2 _____ [2 marks]

3. Gonorrhoea and HIV are both sexually transmitted diseases. Explain why gonorrhoea can be treated with antibiotics, but **not** HIV.

_____

_____ [2 marks]

4. Antibiotic-resistant strains of bacteria are causing problems in hospitals.

Explain the large increase in the number of antibiotic-resistant strains of bacteria.

_____

_____

_____ [4 marks]

Making and testing new drugs

1. Suggest who knows which patients have been given the active drug during a double-blind trial.

_____ [1 mark]

2. Four drugs were compared for their effectiveness in reducing the risk of heart disease. The same number of participants took each type of drug for ten years. No participants had heart disease at the start of the trial.

The results are shown in the graph.

Maths **a** How many more participants taking drug **A** developed heart disease during the trial than those who took drug **D**?

_____ [1 mark]

Maths **b** Comparing the drugs in the trial, how many times better is the most effective drug at reducing heart disease than the least effective drug?

_____ [1 mark]

c In this trial there were the same number of participants taking each drug. This is **not** always the case in drug trials. How could you compare results if there were different numbers taking each drug?

_____ [1 mark]

3. On average it takes ten years to develop a new drug. Describe the different stages in testing that a new drug must go through.

Literacy

Set your answer out clearly to gain marks through a detailed, logical and coherent answer.

_____

_____

_____

_____

_____ [6 marks]

Monoclonal antibodies

1. The diagram shows some stages in the production of monoclonal antibodies.

Higher Tier only

antigens injected   myeloma cells

mouse   B lymphocytes   monoclonal antibodies

a Explain why antigens are injected into the mouse.

_____ [1 mark]

b Tumour cells are combined with lymphocytes to make hybridomas. Explain why.

_____

_____ [2 marks]

2. Gonorrhoea responds to the antibiotic penicillin. Chlamydia is another sexually transmitted disease that produces similar symptoms to gonorrhoea. Diagnosis before treatment is important because chlamydia does **not** respond to penicillin. Adding monoclonal antibodies to a sample taken from the infected region speeds up the diagnosis.

Higher Tier only

a What are monoclonal antibodies?

_____ [1 mark]

b Explain why monoclonal antibodies are useful in diagnosing gonorrhoea.

_____

_____ [2 marks]

c Explain why monoclonal antibodies are **not** as widely used as scientists first hoped.

_____ [1 mark]

3. How can monoclonal antibodies be used in the treatment of cancer?

_____

_____

_____ [3 marks]

Plant diseases

1. A gardener notices purple and black spots on the leaves of her rose plants. Some leaves have turned yellow and dropped off.

a Suggest the disease that may be affecting the rose plant.

_____ [1 mark]

b How does this disease spread?

_____ [1 mark]

c What is the treatment for this disease?

_____

_____ [2 marks]

2. Tobacco mosaic virus (TMV) causes leaf discolouration.

Worked Example

Explain why TMV causes plants to have stunted growth. [4 marks]

TMV causes a lack of nutrition and a low immune system in plants. The virus reproduces in the cells causing growth issues.

Marks gained: [0 marks]

This answer would not gain any marks. The student should have linked the discolouration of the leaves (in the question) with a lack of chlorophyll/ability to photosynthesise. Less photosynthesis means less glucose made; therefore less amino acids/proteins/cellulose for growth; because glucose is needed to make amino acids/proteins/cellulose.

3. Aphids are an example of insect pests that affect plants.

Explain how aphids reduce crop yield.

_____

_____ [2 marks]

Identification of plant diseases

1. Plants need minerals for healthy growth.

a Explain how nitrate deficiency affects plant growth.

_____

_____ [2 marks]

b Plants use active transport to absorb nitrates from the soil.

Describe active transport.

_____

_____ [3 marks]

c Which mineral ion is needed to make chlorophyll?

_____ [1 mark]

d Suggest what a plant lacking in the ion named in **c** above will look like.

_____ [1 mark]

2. Describe how gardeners can identify the pathogen infecting their diseased plants.

_____

_____

_____ [3 marks]

3. Suggest the types of observations needed to detect plant diseases.

_____

_____

_____

_____

_____ [4 marks]

Plant defence responses

1. Plants have physical defences that are barriers to prevent microbial pathogens entering.

Describe physical plant defence responses.

_____

_____

_____ [3 marks]

2. Witch hazel is a plant that produces oil with antibacterial properties.

Explain how this forms part of the plant's defence response.

_____

_____ [2 marks]

3. Passionflower leaves have markings that look like butterfly eggs. This type of mechanical defence is called mimicry.

Suggest how it helps protect the passionflower plant.

_____

_____

_____ [2 marks]

Hint

Examination questions often refer to common varieties of plants that are usually familiar to students in the UK. Sometimes, exotic or unfamiliar species are referred to, in which case, there will usually be some description of the plant to help answer the question.

4. Describe how the nettle plant deters animals.

_____

_____

_____

_____ [3 marks]

Remember

The nettle plant is known for stinging hairs and it is this feature that deters animals. Students are required to show an understanding that the stinging hairs prevent animals from eating the plant and causing damage.

Photosynthesis reaction

1. Why is photosynthesis said to be endothermic?

_____ [1 mark]

2. Write down the word **and** symbol equations for photosynthesis.

_____

_____ [4 marks]

3. Laura investigates the effect of leaf surface area on rate of photosynthesis.

Maths She draws around each leaf as shown.

Calculate the surface area of leaf A and leaf B in cm^3.

> **Maths**
>
> You need to know how to calculate areas of rectangles and triangles for the exam. Since living things are not normally made of these shapes, you might be asked to estimate by counting the number of cm^3 squares covered.

A   B

_____

_____

Surface area of leaf A = _____ cm^3

Surface area of leaf B = _____ cm^3

[2 marks]

4. Suggest how levels of carbon dioxide and oxygen in the atmosphere might be affected by deforestation. Explain this.

Synoptic

_____

_____

_____

_____ [4 marks]

Rate of photosynthesis

1. Antony is provided with the following apparatus:

Required practical

- Canadian pondweed
- Syringe
- Test tube
- Lamp
- Stop clock
- Ruler
- Capillary tube

a Using this apparatus, describe a method Antony could use to investigate the effects of light intensity on the rate of photosynthesis in pondweed.

Your answer should include a detailed method and what variables need to be controlled.

_____

_____

_____

_____

_____

_____

_____

_____

[6 marks]

b Before starting the experiment, Antony cuts each end of the pondweed and adds a drop of 1% sodium hydrogen carbonate to the water surrounding the pondweed.

Explain why he does this.

_____

_____

[2 marks]

2. Jack measures the amount of gas produced by Elodea pondweed by collecting the gas in a glass tube. After 6 minutes the bubble is 35 mm long.

Worked Example

Maths

Calculate the rate of photosynthesis. Give your answer to 2 decimal places, in cm/min. [2 marks]

Rate of photosynthesis = _____ $35 \div 6 = 5.8$ _____ cm/min

Marks gained: [0 marks]

> The question asks for the answer in cm/min but the student has not remembered to convert the 35 mm into cm first. This is an easy mistake to make if you don't read the question carefully. To convert mm into cm the student needed to divide the 35 by 10 which gives 3.5. The second mark is for dividing the distance by the time (3.5÷6) to give the correct answer of 0.58 cm/min.

3. Distance and light intensity are inversely proportional to each other. Explain what this means. [1 mark]

Worked Example

Higher Tier only

The light intensity decreases as the distance increases.

Marks gained: [0 marks]

> This answer is not specific enough so would not get the mark. To gain the mark the student needed to state that the light intensity decreases in proportion to the square of the distance.

4. The inverse square law can be calculated using this formula:

Higher Tier only

$$\text{light intensity} \propto \frac{1}{\text{distance in cm (d)}^2}$$

Maths

Use the inverse square law to calculate the light intensity when the lamp is positioned:

a 10 cm away from the pondweed;

_____

Light intensity = _____
[1 mark]

b 20 cm away from the pondweed.

_____

Light intensity = _____
[1 mark]

> **Maths**
> Inverse proportion just means that as one variable increases, the other variable decreases at the same rate. You will not be expected to know the formula for the inverse square law but you should be able to use it.

5. Suggest a way of measuring light intensity other than by calculating it.

_____ [1 mark]

Limiting factors

1.

Amelie carried out an experiment to find the rate of photosynthesis of a group of aquatic plants at different light intensities.

Higher Tier only

She repeated her experiment at **two** different carbon dioxide concentrations. Her results are shown in the table:

Required practical

Light intensity (arbitrary units)	Rate of photosynthesis (arbitrary units)	
	Low carbon dioxide concentration	**High carbon dioxide concentration**
0.0	0	0
0.2	0.20	0.20
0.4	0.29	0.35
0.6	0.35	0.50
0.8	0.39	0.68
1.0	0.42	0.84
1.2	0.45	0.89
1.4	0.46	0.90
1.6	0.46	0.90

a Plot a graph of Amelie's results.

[4 marks]

Maths

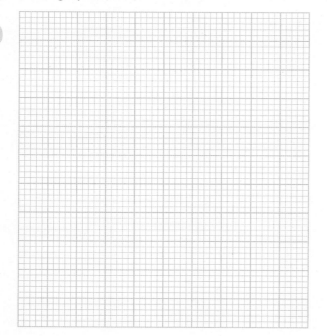

Maths

Make sure you are confident interpreting graphs with two or more limiting factors as these could come up in the higher tier paper.

b Describe and explain the patterns shown by the data.

_____

_____

_____

_____

_____ [4 marks]

The uses of glucose from photosynthesis

1. Which of the following statements are true? Tick **two** boxes.

☐ All plant cells carry out photosynthesis in the day and then switch to respiration at night.

☐ Plant cells can photosynthesise at day or night depending on when they need to produce glucose.

☐ Plant cells respire all the time, but some also carry out photosynthesis when light is available.

☐ Only some plant cells can carry out photosynthesis, some – such as root hair cells – do not. [2 marks]

2. Glucose made in photosynthesis can be converted to other substances for storage. Describe how plants store glucose.

_____

_____ [2 marks]

3. Suggest why it is dangerous for plant cells to build up a high concentration of glucose rather than converting it into other substances.

_____

_____ [2 marks]

4. A farmer wants to know what field he should grow his crop in. He wants the crop to grow well.

The table shows nutrient levels in four of his fields.

Suggest and explain which field would be most suitable for growing his crop in.

Field	% nitrates	% carbonates	% iron oxides
A	0.2	3.6	2.3
B	2.4	2.5	0.6
C	4.1	2.2	2.1
D	0.9	3.3	5.0

_____

_____

_____ [3 marks]

Cell respiration

1. Which statements describe cellular respiration? Tick **two** boxes.

☐ An endothermic reaction that transfers energy to the environment.

☐ An exothermic reaction that transfers energy to the environment.

☐ A series of reactions catalysed by enzymes.

☐ One reaction that occurs in some body cells that need energy. [2 marks]

2. Write down the word **and** symbol equations for aerobic respiration.

_____

_____ [4 marks]

3. Describe **three** ways mammals use energy from respiration.

_____

_____

_____ [3 marks]

4. Metabolism is the sum of all the chemical reactions that take place inside our cells. In some of these reactions, smaller molecules are joined to make bigger ones.

Describe how lipids and proteins are formed from smaller molecules.

Lipids: _____

_____

Proteins: _____

_____ [4 marks]

Anaerobic respiration

1. Explain why our cells normally respire aerobically, rather than anaerobically.

_____

_____ [2 marks]

2. Write down the word equation that represents the chemical reaction involved in making wine.

_____ [2 marks]

3.

Maths

Oliver runs a 400 m race. His blood lactic acid concentration is measured from the start until just after he finishes.

Using evidence from the graph, describe and explain the changes in blood lactic acid concentration during the race.

_____

_____

_____

_____

_____

_____

Maths

If you are asked to 'use evidence from the graph' make sure your answer refers to specific parts of the graph. If the graph allows, pick out data from the graph that supports your point.

_____ [4 marks]

Response to exercise

1. Liam is sprinting away from a bull.

Describe and explain **three** ways his body will respond to the demands of sprinting.

_____

_____

_____ [3 marks]

2. Explain how lactic acid is cleared from the body.

Higher Tier only

_____

_____ [2 marks]

3. The graph shows how oxygen uptake by the lungs changes during exercise and recovery.

Higher Tier only

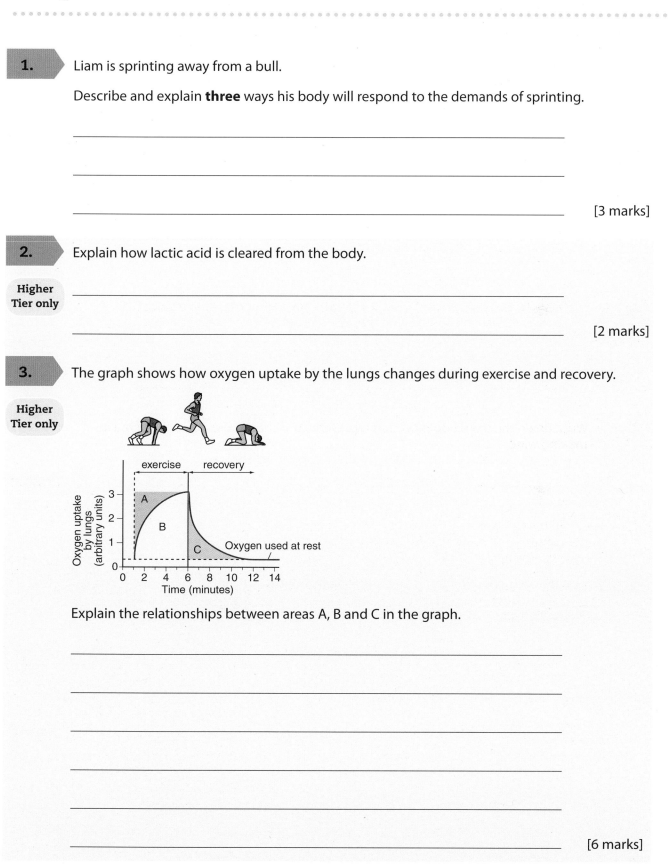

Explain the relationships between areas A, B and C in the graph.

_____

_____

_____

_____

_____

_____ [6 marks]

Homeostasis

1. Define homeostasis.

_____

_____ [2 marks]

2. Name **two** things in the human body that are controlled by homeostasis.

_____

_____ [2 marks]

3. What part of the body monitors water, temperature and carbon dioxide levels in the blood?

_____

_____ [1 mark]

4. Explain why it is important to keep internal conditions constant.

_____

_____ [2 marks]

5. Homeostatic coordination is an automatic response.

Name **two** ways the automatic response may be carried out.

_____

_____

_____ [2 marks]

The nervous system and reflexes

1. Explain how the structure of a neurone is adapted to its function.

_____

_____

_____ [3 marks]

Remember
To gain grades 8 and 9, it is important that you are familiar with the difference between the motor and sensory neurones in terms of whether they conduct impulses to the CNS or away from it.

2. Imagine you could intercept the nerve impulses travelling in the spinal cord. Could you tell which ones came from pain receptors and which from temperature receptors? Explain your answer.

_____

_____

_____ [3 marks]

3. Chloe and Ben were measuring reaction times.

Required practical

- Chloe held a ruler above Ben's hand.

- Chloe let go of the ruler. Ben caught it as quickly as possible.

- They recorded the distance the ruler had fallen.

- They repeated this experiment five more times.

- They worked out the average (mean) result.

a State **two** variables that should be controlled to make the investigation valid.

1 _____

2 _____ [2 marks]

b Chloe and Ben want to investigate the effect of different factors on reaction time. Suggest **two** factors that could affect reaction time.

1 _____

2 _____ [2 marks]

4. Describe the reflex arc.

_____

_____

_____

_____

_____

_____ [6 marks]

The brain

1. The brain is made of which type of cells? _____ [1 mark]

2. Name the parts of the brain that carry out each of the functions below.

Controls unconscious activity such as heartbeat and breathing: _____

Responsible for higher order functions, such as language and memory: _____

Coordinates muscle activity: _____ [3 marks]

3. Explain how scientists have mapped regions of the brain to specific functions.

Higher
Tier only _____

_____

_____

_____

_____ [6 marks]

4. A woman has had a stroke and finds it difficult to speak.

Higher Tier only

a Suggest which part of her brain has been damaged.

_____ [1 mark]

b Her doctors want to try to confirm which part of the brain has been damaged. Suggest **two** ways they could do this.

1 _____

2 _____ [2 marks]

c Explain why it can be difficult to identify which parts of the brain have been damaged.

_____

_____ [2 marks]

The eye

1. This is a diagram of the eye.

a Label the diagram with the following parts.

[5 marks]

| optic nerve | ciliary muscle | retina |
| cornea | suspensory ligaments | |

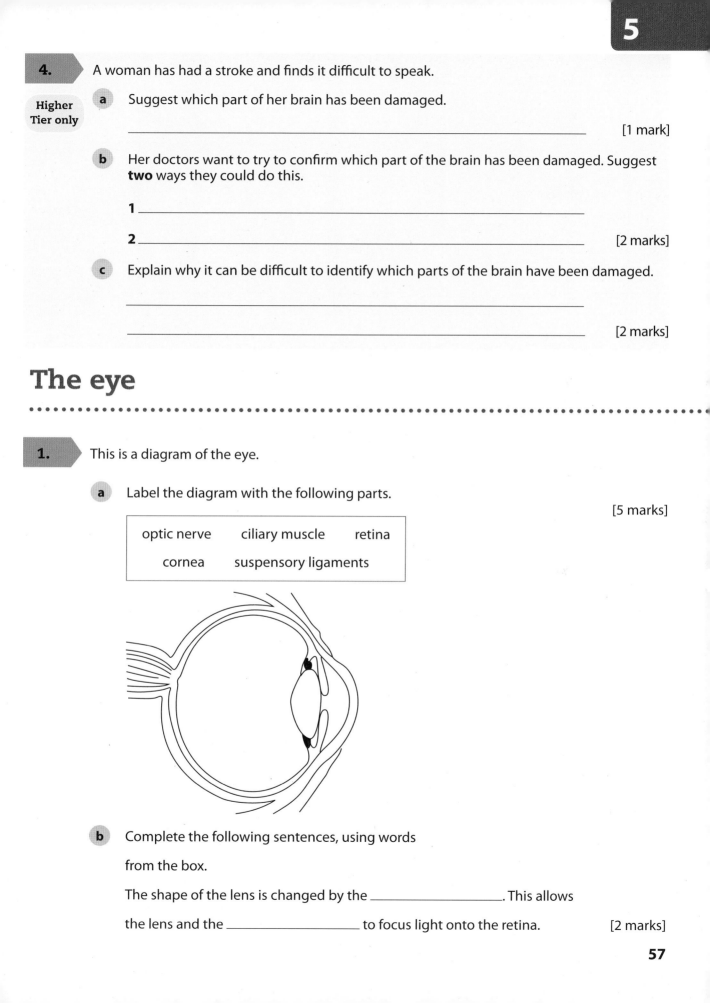

b Complete the following sentences, using words from the box.

The shape of the lens is changed by the _____. This allows

the lens and the _____ to focus light onto the retina. [2 marks]

2. Describe the changes that occur in the eye when light is dim.

_____

_____

_____

_____ [4 marks]

3. Which structures in the human eye help bend light onto receptor cells to produce a sharp focus?

_____ [v2 marks]

4. Describe the changes taking place in the eye when:

a focusing on a **near** object;

_____

_____

_____ [3 marks]

b focusing on a **distant** object.

_____

_____

_____ [3 marks]

5. The diagram shows an eye of a person with short sight (myopia) looking at a distant object.

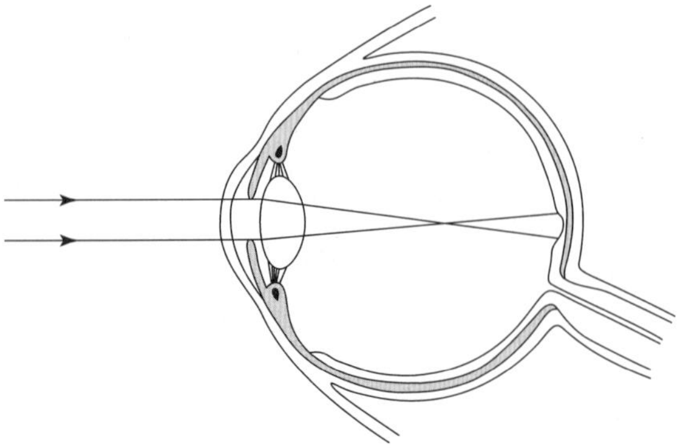

Draw a lens in front of the eye that can correct short sight.

Draw rays of light on the diagram to show how the new lens corrects short sight. [3 marks]

6. Wearing glasses is one way to correct defective eyesight.

Give **two** other ways to correct defective eyesight.

1 _____

2 _____   [2 marks]

Control of body temperature

1. Each student in a class measured their own body temperature.

Their results are shown in the bar chart.

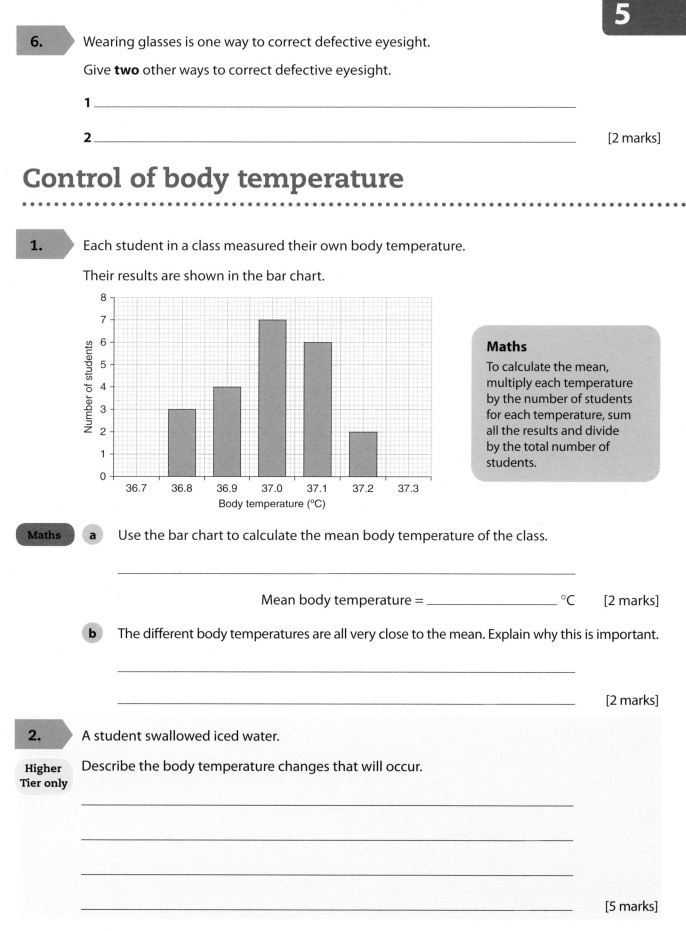

Maths

To calculate the mean, multiply each temperature by the number of students for each temperature, sum all the results and divide by the total number of students.

Maths **a** Use the bar chart to calculate the mean body temperature of the class.

_____

Mean body temperature = _____ °C   [2 marks]

b The different body temperatures are all very close to the mean. Explain why this is important.

_____

_____   [2 marks]

2. A student swallowed iced water.

Higher Tier only Describe the body temperature changes that will occur.

_____

_____

_____

_____   [5 marks]

Hormones and the endocrine system

● ●

1. Label the diagram with the names of the endocrine glands indicated (numbered 1–6).

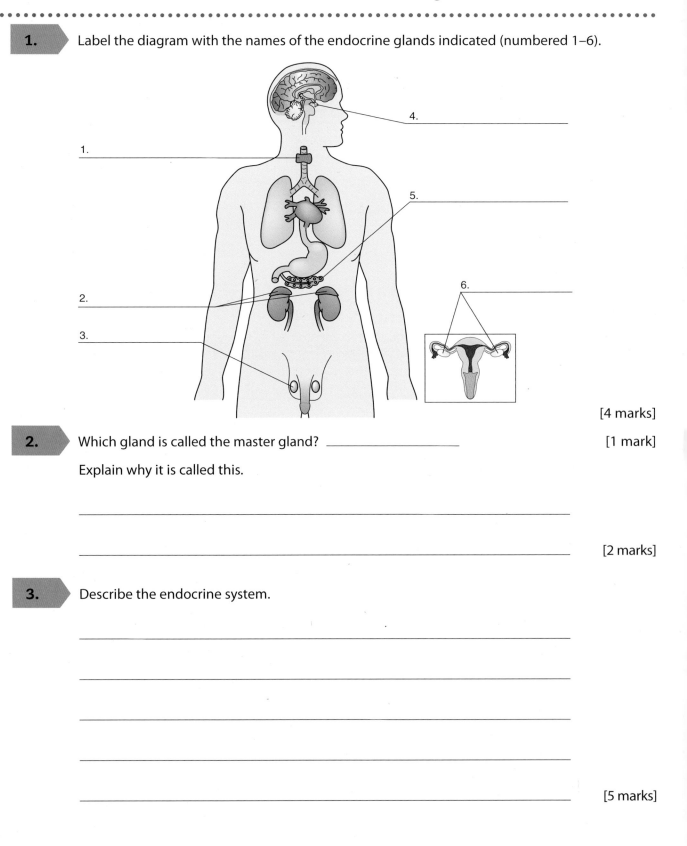

[4 marks]

2. Which gland is called the master gland? _____ [1 mark]

Explain why it is called this.

_____

_____ [2 marks]

3. Describe the endocrine system.

_____

_____

_____

_____

_____ [5 marks]

Controlling blood glucose

1. Complete the following sentences.

The concentration of glucose in the blood is monitored and controlled by the

_____ .

If the blood glucose concentration is too high, the hormone _____ is released into the blood.

This hormone causes glucose to move from the blood into _____

and _____ cells.

In these cells, excess glucose is converted into _____ and stored.   [5 marks]

2. Bread is a source of starch. When we eat bread the starch is broken down into glucose, which enters the blood.

The graph shows the effects on blood glucose concentration of eating different types of bread. The amount of bread eaten was the same in each case.

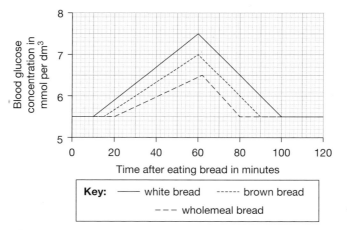

Explain which type of bread would be most suitable for a person with diabetes.

_____

_____

_____

_____

_____   [4 marks]

3. Describe the sequence of events that occurs when blood glucose concentration is too low. [3 marks]

When blood glucose is low glycogen converts glucagon to glucose.
The glucose goes into the bloodstream.

Marks gained: [1 mark]

This answer gained 1 mark for the idea that glucose enters the bloodstream. However, no credit is given to the other part as the student has confused glycogen and glucagon. Remember that **glucagon** is the hormone and **glycogen** is the form of storage of glucose in animal calls. Another mark could have been gained if the student had mentioned the role of the pancreas in secreting glucagon to act on the liver to convert glycogen to glucose.

Maintaining water and nitrogen balance in the body

1. Name **three** ways in which water is lost from the body.

_____

_____ [3 marks]

2. **a** What is the hormone that controls water excretion?

_____ [1 mark]

b How does it work?

_____ [1 mark]

3. Describe how excess amino acids are processed by the liver before being excreted by the kidney.

_____

_____

_____

_____ [4 marks]

4. Describe how the amount of water in the bloodstream is controlled by the kidneys.   [3 marks]

Higher Tier only

After filtering the blood, the kidneys reabsorb water to keep water levels constant in the blood. More water is reabsorbed in the presence of the hormone ADH, so concentrated urine is produced, and less water is reabsorbed if there is a low level of ADH in the bloodstream.

Marks gained: [3 marks]

This answer gains full marks. The student has given an accurate description of how water levels are controlled by the kidneys.

Hormones in human reproduction

1. **a** Name the female reproductive hormone and the organ where it is produced.

_____   [1 mark]

b Name the male reproductive hormone and the organ where it is produced.

_____   [1 mark]

c What is the role of testosterone?

_____   [1 mark]

2. **a** Where is FSH produced?

_____   [1 mark]

b What is the target organ for FSH?

_____   [1 mark]

c How does FSH move from where it is produced to its target organ?

_____   [1 mark]

3. Describe **one** role for each hormone in the menstrual cycle.

FSH: _____

LH: _____ [2 marks]

4. The menstrual cycle is controlled by several hormones. The graphs show concentrations of four hormones at different times during the menstrual cycle.

Higher Tier only

Maths

When interpreting graphs like these, make sure you use information from both graphs presented. You must relate the peak in oestrogen levels to LH production and the peak of LH production to ovulation.

Use these graphs and your knowledge of reproductive hormones to explain:

a how the concentration of oestrogen affects and controls the development and release of an egg during the cycle;

_____

_____

_____ [3 marks]

b why progesterone continues to be produced throughout pregnancy.

_____

_____

_____

_____ [3 marks]

Contraception

1. Explain the advantages of using birth control pills.

_____

_____

_____ [3 marks]

> **Remember**
>
> You need to learn and understand the action of oral contraceptive pills as well as the advantages and disadvantages of this method of contraception.

2. Read the information.

> The development of the contraceptive pill involved human trials. A large trial took place in Puerto Rico in 1956. The women taking part were given a pill containing a high dose of hormones. They were told that the pill would prevent pregnancy. They were **not** told that the drug was experimental and that they were taking part in a trial. They also were **not** told about possible side effects. The side effects for some women included nausea, depression and weight gain. Three women died due to blood clotting that may have been caused by the trial.
>
> After the trial showed that the pill worked successfully as a contraceptive, it was made available to the public in the United States and other countries around the world. The contraceptive pills used today contain a much lower dose of hormones than those used in the trial.

Evaluate the trial in Puerto Rico.

_____

_____

_____

_____

_____

_____

_____

_____ [6 marks]

Using hormones to treat human infertility

1.   Which hormones can be used to treat infertility?

Higher Tier only
_____ [2 marks]

2.   There are some advantages and disadvantages of using fertility and contraceptive drugs.

Higher Tier only
a   Explain **two** disadvantages of using contraceptive drugs on fertility.

_____

_____

_____ [2 marks]

b   Explain **two** other disadvantages of using contraceptive drugs.

_____

_____

_____ [2 marks]

3.   Describe the process of _in-vitro_ fertilisation (IVF) treatment.

Higher Tier only
_____

_____

_____

_____

> **Remember**
> Make sure you understand that FSH and LH are two separate hormones with two very specific functions. Make sure you explain the importance of the FSH injection in causing the formation of several eggs.

[4 marks]

4. Jane is 48 and John is 50. They both want to try IVF to have a child.

Higher Tier only What are the drawbacks of having IVF at this age?

_____

_____

_____

_____ [3 marks]

Negative feedback

1. Explain how insulin controls blood glucose concentration by negative feedback.

Higher Tier only

_____

_____

_____

_____ [4 marks]

2. What is the role of adrenaline in the body?

Higher Tier only

_____

_____

_____

_____ [4 marks]

3. **a** Where is thyroxine produced? _____ [1 mark]

Higher Tier only **b** Name **two** functions of thyroxine.

_____

_____ [2 marks]

Plant hormones

1. Plants respond to different environmental factors because of hormones.

a Which environmental conditions affect direction of root growth?

_____ [2 marks]

b Which hormone is responsible for the direction of growth of roots?

_____ [1 mark]

c Explain how this hormone is able to influence direction of growth of roots.

_____

_____ [2 marks]

2. Anna investigated phototropism in plant shoots.

Required practical

Shoot **A** had a thin piece of plastic inserted just below the tip.

Shoot **B** was **not** treated in this way.

Both shoots were left for several days as shown in the diagram.

Light from one side only — Plastic — Shoot **A**

Light from one side only — Shoot **B**

a Draw the two shoots after several days.

_____ [2 marks]

b Explain your answers to part **a**.

_____

_____

_____

_____ [4 marks]

Sexual and asexual reproduction

1. The offspring of a zebra and a horse is called a zorse.

James thinks a zorse is the product of cloning.

Jane says it is sexual reproduction.

Jade says it is an example of asexual reproduction.

a Who is correct? _____ [1 mark]

b Explain why.

_____

_____

_____ [3 marks]

2. Which of the following do **not** play a part in asexual reproduction? Tick **more than two** boxes.

☐ Mitosis ☐ Gametes

☐ Meiosis ☐ Cell division

☐ Chromosomes ☐ Zygote [2 marks]

Literacy

The marks for some multiple-choice questions may not reflect the number of answers required. For example, here there are two marks available but three correct answers. These questions require a thorough understanding of the topic.

3. **a** Describe the main ways that asexual reproduction is different from sexual reproduction.

_____

_____

_____

_____ [4 marks]

b Some species, such as the malarial parasite, can reproduce both asexually and sexually.

Suggest and explain **one** factor that would favour sexual reproduction.

_____

_____ [2 marks]

4. Explain the advantages of asexual reproduction in plants.

_____

_____

_____

_____

_____

_____ [6 marks]

Cell division by meiosis

1. Both sex cells and body cells are produced by cell division.

a State the type of cell division that occurs **only** in sex cell formation.

_____ [1 mark]

b Explain why mutations in sex cells may be **more serious** than mutations in body cells.

_____

_____ [1 mark]

2. **a** Name the gametes produced by plants.

_____ [1 mark]

b Name the gametes produced by animals.

_____ [1 mark]

c State the number of chromosomes in:

a normal human cell; _____

a human gamete; _____

the daughter cell from mitosis of a human cell. _____ [3 marks]

3. Compare the processes of meiosis and mitosis. Mention where each takes place and the kind of products made.

_____

_____

_____

_____

_____

_____ [6 marks]

DNA, genes and the genome

1. The table shows the number of chromosomes in body cells and gametes for four different organisms. There is a fixed number of gametes in each species.

Complete the table.

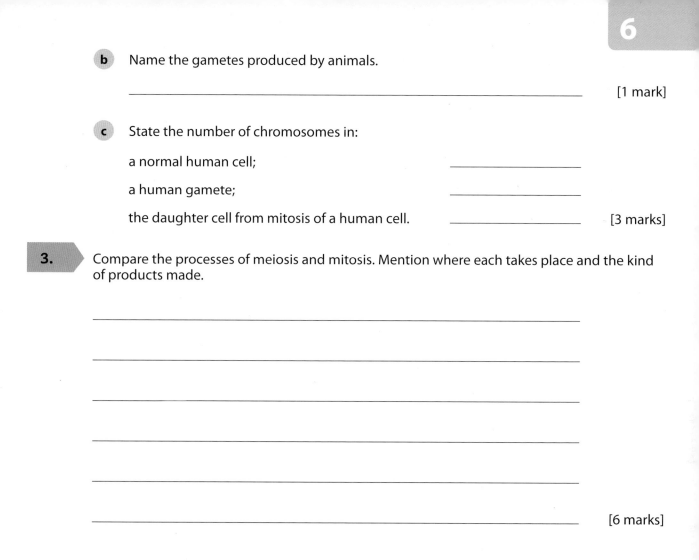

Organism	Chromosome number in body cells	Chromosome number in gametes
Human	46	_____
Mouse	_____	20
Garden pea	14	_____
Sloth	420	_____

Remember
You may be given information about an organism and asked to work out chromosome numbers from this. Remember that chromosomes of gametes are always half the number of body cell chromosomes.

[2 marks]

2. Describe what is meant by the term genome.

_____ [1 mark]

3. Understanding the human genome will help scientists look for genes linked to different diseases.

a Understanding the human genome could help to predict if someone will get cancer.

Suggest **two** reasons why some people might object to this use.

1 _____

2 _____ [2 marks]

b Give **two** other reasons why it is important to study the human genome.

1 _____

2 _____ [2 marks]

4. A gene is a small section of DNA on a chromosome.

Synoptic Explain the connection between genes, proteins and cell structure.

_____

_____

_____

_____

_____

_____

_____

_____

_____

_____

_____ [6 marks]

Structure of DNA

1. Describe the structure of a nucleotide.

_____

_____ [3 marks]

2. The DNA molecule is made up of repeating nucleotide units.

a How are the nucleotides joined? _____ [1 mark]

b The sequence of bases forms a code to instruct the cell to make a particular protein.

How many bases form the code for **one** amino acid? _____ [1 mark]

Higher Tier only **c** Suggest what would happen if the sequence of bases was changed.

_____

_____ [2 marks]

3. **Not all parts** of the DNA code for proteins.

Higher Tier only **a** Explain the function of the non-coding parts of DNA.

_____

_____ [1 mark]

b Suggest what would happen if the genetic code is changed in the non-coding parts of DNA.

_____

_____ [1 mark]

4. A DNA molecule consists of two strands coiled to form a double helix.

Higher Tier only **a** Chemical analysis of DNA molecules, from a variety of cells, shows that the total number of molecules of A (adenine) plus G (guanine) equals the total amount of T (thymine) plus C (cytosine).

Suggest a conclusion which can be made from this observation.

_____

_____ [2 marks]

b Explain how a gene codes for the production of a protein.

_____

_____

_____

_____ [4 marks]

Protein synthesis and mutations

1. The diagram shows the bases on one strand of a section of DNA.

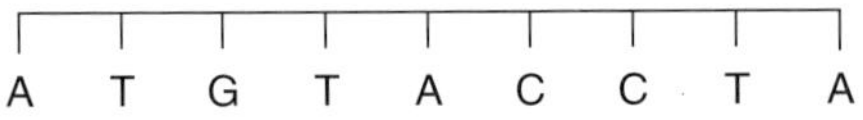

A   T   G   T   A   C   C   T   A

Higher Tier only **a** Write out the sequence of bases on the complementary strand.

_____ [2 marks]

b State how many amino acids are coded for by the strand of DNA.

_____ [1 mark]

2. Amino acids join together during protein synthesis. Tick **one** box.

Amino acids are joined together:

☐ at the membrane   ☐ in the mitochondria

☐ in the nucleus   ☐ at the ribosomes   [1 mark]

3. Describe how a protein is produced from a strand of DNA.

Higher Tier only

_____

_____

_____

_____

_____

_____

_____

_____   [6 marks]

4. DNA codes for the sequence of amino acids making up an enzyme.

Synoptic **a** Explain the role of the active site of an enzyme.

_____

_____   [2 marks]

Worked Example **b** Explain how a mutation in the DNA coding for an enzyme could affect
its active site.   [3 marks]

Synoptic

Higher Tier only

*A mutation is a change in the sequence of base
pairs in a gene. It could cause a change in the
amino acids that join together to form the protein.
If the protein is an enzyme, this could change the
shape of the active site so the enzyme might not
work properly as it wouldn't be able to bind to the
substrate.*

This is a well-structured
answer that includes good
scientific knowledge;
it would gain full marks.

Marks gained   [3 marks]

Inherited characteristics

1. Sometimes a pair of spotted leopards will have an offspring with black fur. This is known as a black panther.

Spotted fur is controlled by a dominant allele **B**.

Black fur is controlled by a recessive allele **b**.

a Explain what is meant by the term 'allele'.

_____ [1 mark]

b State the possible genotypes of a spotted leopard and a black panther.

Spotted leopard: _____

Black panther: _____ [3 marks]

Maths

Higher Tier only

c Draw a Punnett square diagram to show the possible offspring between a black panther and a heterozygous spotted leopard.

Label your diagram to show the phenotypes of all the offspring.

State the probability of **one** of the offspring being a black panther.

Probability of offspring being a black panther = _____ [4 marks]

2. *Drosophila* fruit flies have been used in many experiments investigating genetic inheritance.

Maths

In one experiment, pure-bred flies with long wings were bred with pure-bred flies with short wings. All the offspring had long wings.

Higher Tier only

long wing fruit fly   short wing fruit fly

Next, these long wing offspring were bred amongst themselves.

Predict and explain the results of this cross. Draw a Punnett square diagram as part of your answer. Choose suitable symbols for the alleles for long wing and short wing.

[6 marks]

Inherited disorders

1. Polydactyly is an inherited condition causing extra fingers or toes. It is caused by a dominant allele D.

The family tree shows how polydactyly was inherited in Sam's family.

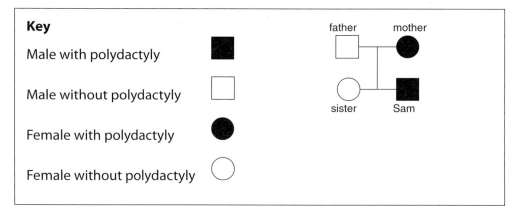

Use the family tree to work out the genotype of each member of Tom's family.

Sam _____

mother _____

father _____

sister _____ [4 marks]

2. Cystic fibrosis is an inherited disorder caused by a recessive allele **f**.

Ali has the genotype **ff**. His wife Shara has the genotype **FF**. They want to have children but are worried about the possibility of their children or grandchildren having cystic fibrosis. They ask their GP about this.

Suggest what the GP would tell them.

_____

_____

_____

_____ [4 marks]

Sex chromosomes

1. The two sex chromosomes in humans are X and Y.

    a State the genotype of a female. _____ [1 mark]

    b State the genotype of a male.

                                                   _____ [1 mark]

2. Steve and Jane have two sons. Jane is pregnant with a third child.

Steve says there is a higher chance that the new baby will be a girl.

Is he right or wrong? _____

Explain why. _____

_____ [2 marks]

3. Thomas and James are both boys because of the sex chromosomes inherited from their parents.

These statements are about the inheritance of sex chromosomes.

Which are true? Tick **more than two** boxes.

☐ Each body cell in a human male contains XX sex chromosomes.

☐ A gene on the X-chromosome determines the sex of the embryo.

☐ The sex-determining gene triggers the development of the ovaries.

☐ A gene on the Y-chromosome determines the sex of the embryo.

☐ Each body cell in a human female contains XX sex chromosomes.

☐ The sex-determining gene triggers the development of testes. [2 marks]

4. Henry VIII was a king of England in the sixteenth century. He blamed several of his wives for giving birth to daughters and not sons. We now know that the sex of a baby is determined by the father, not the mother.

Explain why this is.

_____

_____ [2 marks]

Variation and mutations

1. What is meant by the term variation?

_____ [1 mark]

2. Give **one** environmental factor and **one** other factor that might affect the height of plants growing near a wall.

Environmental factor: _____ [1 mark]

Other factor: _____ [1 mark]

3. In a population of bacteria, there is a mutation in one bacterial cell that makes it resistant to a particular antibiotic.

Explain how eventually the whole population may become resistant.

_____

_____

_____ [3 marks]

4. Iguanas are reptiles that usually have green or yellow skin.

Some iguanas on the Galapagos Islands have pink skin.

a The colour of the pink iguana is the result of a gene mutation.

Which statement about gene mutations is true? Tick **one** box.

1 Gene mutations change the base sequence of DNA.

2 Gene mutations are always harmful.

☐ 1 only ☐ 2 only ☐ Both 1 and 2 ☐ Neither 1 nor 2 [1 mark]

Synoptic **b** The change in the DNA sequence that causes the pink skin results from alteration of four amino acids.

How many bases code for four amino acids? _____ [1 mark]

Synoptic **c** DNA analysis suggests that the pink iguanas split from the more common yellow iguana population 5.7 million years ago.

Give **one** reason why the fossil record has not been able to support this data.

_____ [1 mark]

Theory of evolution

1. Explain what is meant by the term 'evolution'.

_____

_____ [2 marks]

2. **a** Charles Darwin and Alfred Russel Wallace developed the theory of natural selection to explain how evolution happens.

Explain the process of natural selection.

_____

_____

_____

_____

_____

_____ [5 marks]

b An earlier theory to explain evolution had been suggested by Jean-Baptiste Lamarck.

Describe Lamarck's theory.

_____

_____ [1 mark]

c Explain why Lamarck's theory was later discredited.

_____

_____ [2 marks]

3. The colour of grove snail shells varies from a very pale yellow to dark brown.

Snails with paler shells are more common in grassland. Snails with darker shells are more common in woodland.

Predators of grove snails include birds like the song thrush.

Suggest an explanation for the distribution of the differently coloured snails.

_____

_____

_____

_____ [4 marks]

4. Fennec foxes live in the Sahara desert. They have very large ears to lose excess heat. Arctic foxes have very small ears.

Worked Example

Explain how natural selection could have caused Arctic foxes to have small ears. [6 marks]

The ancestors of Arctic foxes had larger ears on average than modern Arctic foxes. However, there was variation in the size of their ears. Foxes with smaller ears lost less heat and so were better able to survive in cold conditions than foxes with larger ears. Therefore the foxes with smaller ears were more likely to breed and pass on the alleles for smaller ears to their offspring. Over time, therefore, the average ear size got smaller.

Marks gained: [6 marks]

This is a very clear explanation that gains full marks. Although the examples may change, the stages in natural selection are always the same: variation, survival of those best adapted to the environment ('survival of the fittest'), breeding of those best adapted and passing on beneficial characteristics to their offspring.

Speciation

1. The Galapagos Islands lie off the coast of South America. Different species of finch, a type of bird, live on the different islands. The different species are different sizes and have differently-shaped beaks.

They are all similar to another species of finch that lives on the mainland.

a Explain how the different species of finch on the islands could have evolved.

_____

_____

_____

_____

_____ [4 marks]

b The Galapagos finches are sometimes known as Darwin's finches because Charles Darwin studied them as he developed his theory of evolution by natural selection.

Explain why Darwin's theory took a while to become widely accepted.

_____

_____

_____

_____ [3 marks]

2. The Amazon rainforest is home to one in ten of the known species on Earth. One theory to help explain the great number of species is that over many thousands of years the size of the rainforest has decreased and increased many times due to climate change. During drier periods the forest has shrunk into smaller individual forests, separated by open land. During wetter periods, the forested areas have increased in size joining together again.

Suggest how changes in the size of the rainforest could have given rise to so many species.

_____

_____

_____

_____

_____

_____

_____

_____ [6 marks]

The understanding of genetics

1. Mendel crossed pure breeding tall plants with pure breeding short plants. All offspring were tall.

a What are pure breeding tall plants?

_____ [1 mark]

Maths **b** Step 1: Mendel crossed offspring from this experiment and collected the seeds.

Step 2: He planted the seeds and measured the height of the plants (trial 1).

He repeated both steps to give further data (trials 2 and 3):

Trial	Total number of seeds planted	Number of tall pea plants	Number of short pea plants
1	100	76	24
2	100	75	25
3	100	74	26

Do you think data from the trials is representative? Explain your reasoning.

_____

_____ [2 marks]

2. Mendel used the same process to investigate how other characteristics were inherited in pea plants.

He crossed the offspring produced from parent plants that were pure breeding for each characteristic. He used 100 seeds in each trial.

This table shows his results.

Characteristic	Results
Colour of flower	76 purple 24 white
Colour of seed	74 yellow 26 green
Colour of seed pod	75 green 25 yellow

a Give **two** characteristics of pea plants that are dominant.

_____ [2 marks]

Maths **b** Mendel used the same process in another trial.

150 of the pea plants produced green seeds. Predict the number producing yellow seeds. Show your working.

_____

Number producing yellow seeds = _____ [2 marks]

c What plants need to be crossed to produce only seeds that will grow into short white-flowered plants? Explain your reasoning.

_____

_____ [2 marks]

3. The scientific work by many scientists led to the development of gene theory.

Explain the advantages of using **teams** of scientists to investigate scientific problems.

_____

_____

_____

_____ [3 marks]

Fossil evidence

1. Scientists have found fossils of a 'giant' penguin called Icadyptes.

a A study of fossils gives evidence for which theory?

_____ [1 mark]

b Suggest the types of fossil found.

_____

_____ [2 marks]

c Why are scientists **not** certain that modern penguins evolved from Icadyptes?

_____

_____ [2 marks]

2. Suggest **three** reasons why the fossil record is incomplete.

1 _____

2 _____

3 _____ [3 marks]

> **Remember**
>
> It is important to understand why few organisms have been found as fossils.
>
> For example, some geological processes are so great that they fold rocks over so older rocks are sometimes found above younger rocks. You should be able to explain the impact of examples like this on the challenges of interpreting how species have evolved from fossil evidence.

Other evidence for evolution

1. Which reasons, when put together, can cause antibiotic resistance?

Tick **two** boxes.

☐ Increased use of antibiotics.

☐ Random changes in the genes of microorganisms.

☐ Increased use of disinfectants in hospitals.

☐ Increased use of vaccines.

☐ People always finishing a course of antibiotics.

☐ Development of new antibiotics. [2 marks]

2. Most antibiotics are made in pharmaceutical factories, although they have often first been discovered in nature.

Synoptic Suggest why discovering new antibiotics may become more difficult in the future.

Worked Example

Biodiversity is reducing as rainforests are being destroyed by deforestation so there are fewer places to get new drugs from.

> This is a good answer to this synoptic question. Other acceptable reasons for habitats being damaged include mining and pollution.

Marks gained: [2 marks]

3. **a** Explain how antibiotic resistance evolves in bacteria.

_____

_____

_____

_____

_____

_____ [6 marks]

b Give **two** ways to reduce the rate of development of antibiotic-resistant bacteria.

1 _____

2 _____ [2 marks]

Extinction

1. Mammoths were related to modern day elephants. The last mammoths went extinct about 4000 years ago.

a What is meant by the term 'extinct'?

_____ [1 mark]

b Describe the factors that contribute to extinction of a species.

_____

_____

_____

_____ [4 marks]

Remember
When answering a question on extinction, use the word 'change'. If there is no change then species do **not** become extinct.

c   Suggest **two** reasons scientists think may have contributed to the extinction of the mammoth.

_____

_____   [2 marks]

Selective breeding

1.   Domesticated dogs are descended from wild wolves.

All modern breeds of dog are the result of selective breeding over thousands of years.

Suggest **one** characteristic that the first dogs were selected for and explain why.

Characteristic: _____

Explanation: _____   [2 marks]

2.   A farmer wants to produce better apple trees using selective breeding. He wants to breed two apple trees together:

- apple tree **A** which produces very sweet small apples

- apple tree **B** which produces large apples which are not so tasty.

He uses the following method.

- As soon as they appear, cover the flowers from apple tree **A** with cotton bags.

- Use a small brush to collect pollen from the flowers of apple tree **B**.

- Remove one of the bags from apple tree **A**, brush the flowers with the pollen from apple tree **B**, replace the bag over the flower.

- Repeat the previous steps until all the flowers from apple tree **A** have been pollinated with pollen from apple tree **B**.

- When apple tree **A** produces fruit, collect the seeds from the apples.

a   Suggest what characteristics the farmer wants in the new apple trees he is going to produce.

_____   [1 mark]

b Can the farmer be sure of the characteristics he will get in the new apple trees?

Explain your answer.

_____

_____ [1 mark]

c Suggest why the farmer kept the flowers of apple tree **A** covered with cotton bags.

_____

_____ [1 mark]

d After the farmer collected the seeds from apple tree **A**, describe the next steps in the process.

_____

_____

Hint
The basic steps of selective breeding are the same whatever species you are working with.

_____

_____

_____ [4 marks]

3. A farmer has selectively bred tomato plants to grow more quickly.

a Suggest how the selectively-bred tomato plants may increase the farmer's profits.

_____ [1 mark]

b Explain how selective breeding affects the genetic variation of the farmer's tomato plants.

_____

_____

_____ [3 marks]

Genetic engineering

1. Explain the term 'genetic engineering'.

_____ [1 mark]

2. The diagram shows a technique used in genetic engineering.

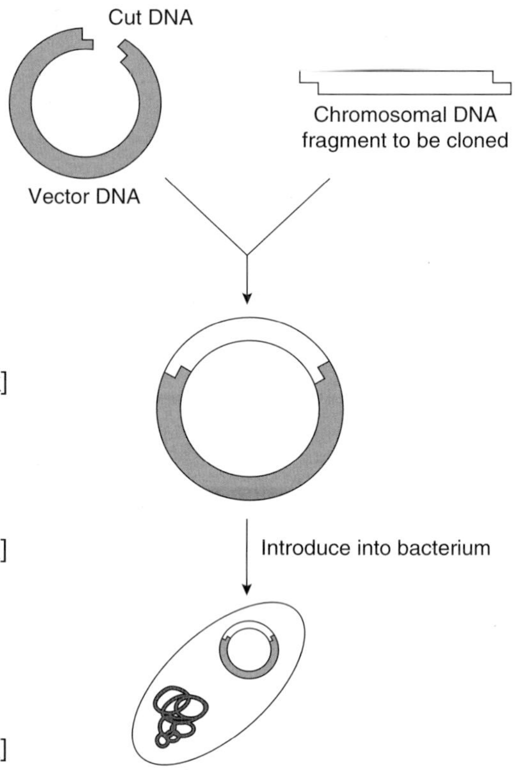

a Why is a vector used?

_____

_____

_____ [1 mark]

b Name the vector.

_____ [1 mark]

c Suggest how the DNA of the vector and chromosome is cut.

_____ [1 mark]

d Suggest how the DNA of the vector and chromosome is joined together.

_____ [1 mark]

Cut DNA

Vector DNA

Chromosomal DNA fragment to be cloned

Introduce into bacterium

3. Cauliflowers were genetically engineered to make their leaves greener.

'Green genes' were introduced into cauliflower cells. The 'green genes' were attached to part of a cauliflower virus or to a bacterial plasmid.

a How is it possible to use a cauliflower virus without harming the cauliflower cells?

_____ [1 mark]

b Suggest **two** concerns about this genetic engineering.

_____

_____

_____ [3 marks]

4. Natural insulin can be taken from the pancreas of pigs or cows.

Suggest advantages of using genetically engineered bacteria to produce insulin.

_____

_____

_____ [3 marks]

Cloning

1. One method of cloning is 'embryo transfer'. This is sometimes used when selectively breeding cattle.

a Write numbers in the empty boxes to show the order of the stages in embryo transfer.

Stage	Order
Collect sperm from selected bull and eggs from selected cow	
Cells of embryo split apart forming separate embryos	
Each embryo implanted into a surrogate cow	
Fertilisation using IVF	
Fertilised egg grows into an embryo	

[2 marks]

Synoptic **b** Explain the advantage of this method over traditional selective breeding.

_____

_____

_____ [2 marks]

2. Scientists have produced a disease-resistant tomato plant using genetic engineering techniques. Now they want to quickly produce many copies of the disease-resistant plant. They have three options:

A – collect seeds from the plant and use these to grow new plants

B – take cuttings from the plant

C – use tissue culture.

a Which option should the scientists choose? _____ [1 mark]

b Explain your answer to part **a**.

_____

_____

_____ [3 marks]

3. Scientists plan to use adult cell cloning to recreate a living mammoth.

They will use a skin cell from the body of a frozen mammoth, and an unfertilised egg cell from a living elephant.

a Describe how the scientists could use adult cell cloning to clone the mammoth.

_____

_____

_____

_____ [4 marks]

b Explain why the scientists expect the animal produced to look like a mammoth and **not** an elephant.

_____ [1 mark]

6

Classification

1. **a** What is meant by the term 'classification'?

_____ [1 mark]

b Name the Swedish scientist who developed the classification system for living organisms that we still use today.

_____ [1 mark]

c The Swedish scientist also developed the binomial system for naming species.

Explain what is meant by the 'binomial system'.

_____

_____ [2 marks]

Synoptic **d** What is a species?

_____

_____ [2 marks]

2. In 1977, Carl Woese suggested that living organisms are classified into a 'three-domain' system.

One domain is the eukaryota which consists of all eukaryotes. The other two domains consist of prokaryotes.

a Name **two** kingdoms that are part of the eukaryota.

_____

_____ [2 marks]

b Name the **two** domains that consist of prokaryotes.

_____

_____ [2 marks]

c What technology did Woese use to construct the three domains that was **not** available to the Swedish scientist?

_____ [1 mark]

Habitats and ecosystems

1. Explain the relationship between a habitat, a community and an ecosystem.

_____

_____

_____ [3 marks]

2. Duckweed, an aquatic plant, grows in a tank of water. The number of plants is counted every five days. The results are shown below:

Maths

Day	0	5	10	15	20	25	30	35	40
Number of duckweed plants	1	2	4	8	16	32	31	31	31

a Plot the results on an appropriate type of graph to show how the population size changes with time. [4 marks]

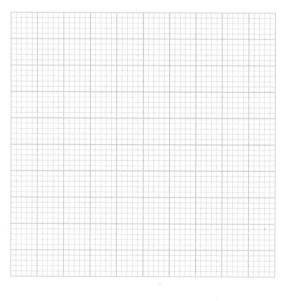

Maths

Make sure you know what type of graph to use for different types of data. Put the independent variable on the x-axis and the dependent variable on the y-axis. Only add a line of best fit if you have to.

b Describe and explain the shape of the graph.

_____

_____

_____ [3 marks]

Food in the ecosystem

1. The diagram shows a food chain.

phytoplankton ⟶ zooplankton ⟶ minnow ⟶ pike

a Which term describes the position of zooplankton in the food chain?

_____ [1 mark]

b What do the arrows represent?

_____ [1 mark]

2. Food webs are often used rather than food chains to show feeding relationships between organisms.

Explain the advantages of representing feeding relationships with a food web rather than a food chain.

_____

_____ [2 marks]

3. Explain why it is rare to have more than five trophic levels in an ecosystem.

_____

_____ [2 marks]

4. For an ecosystem to be stable and self-supporting, it must have an external source of energy. This is usually the Sun or an artificial light source.

Explain why this is.

_____

_____

_____

_____ [4 marks]

Abiotic and biotic factors

1. List **four** abiotic factors that could affect the distribution and abundance of vegetation on a north-facing compared with a south-facing wall.

_____

_____ [4 marks]

2. An invasive species can be any kind of living organism that is **not** native to an ecosystem and which causes harm. Invasive species cause a decline in native wildlife populations.

Describe **three** ways that an invasive species could cause decline of a native species.

_____

_____

_____ [3 marks]

3. The graph shows a predator–prey relationship. It shows annual changes in the populations of two different mites, one of which preys upon the other.

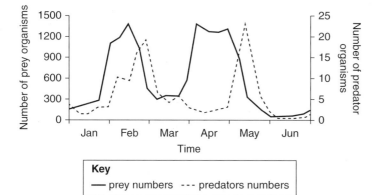

Key
— prey numbers - - - predators numbers

Remember

You might be asked to interpret a predator–prey graph in the exam. Make sure you can explain why they are out of phase with each other.

Explain why the predator curve of the graph is **out of phase** with the prey curve.

_____

_____

_____ [3 marks]

4. Two species of barnacles are found living close together on the rocky shores in the intertidal zone along the British coastline. The intertidal zone is the area of land where the tide meets the sea.

Chthamalus barnacles generally occupy a zone higher up the shore than *Balanus* barnacles.

The diagram shows where these barnacles are found.

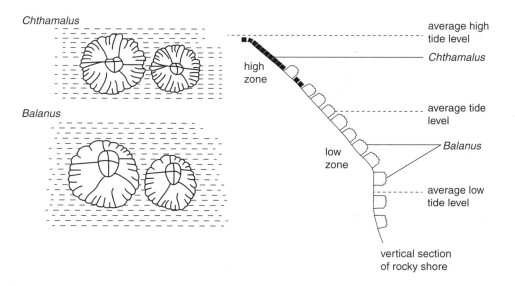

a Suggest and explain a reason why *Chthamalus* barnacles are found higher up the shoreline.

_____

_____

_____

_____ [3 marks]

b *Chthamalus* can actually live in the low zone, but only if the zone is free from *Balanus*.

Suggest why *Chthamalus* are **not usually found** in the low zone.

_____

_____

_____ [2 marks]

Adapting for survival

1. Describe the **difference** between structural, behavioural and functional adaptations.

_____

_____

_____ [3 marks]

2. Silver ants are extremophiles living in the Sahara Desert. They come out at the hottest point in the day, when temperatures are around 60 °C. The ants scavenge for the corpses of insects which have died of heat exposure.

a Describe what is meant by the term 'extremophile'.

_____ [1 mark]

b Suggest **another reason** why coming out at the hottest time could be a behavioural adaptation.

_____ [1 mark]

3. Hydrophytes are flowering plants adapted to living in water. They often have flexible stems and leaves along with roots projecting into the air. Suggest and explain how these features help the plant survive.

_____

_____ [2 marks]

4. Describe and explain the adaptions that help a human to maintain a consistent body temperature. [6 marks]

Worked Example

Synoptic

If the body temperature gets too high the hairs on the skin lie flat so less air is trapped near the surface of the skin which means they aren't insulating air kept close to your skin and the heat can be more easily transferred to the environment.

We have sweat glands which help us stay cool because as the sweat evaporates away from

> This answer is logically ordered, starting with how the body adapts to colder conditions and then moving on to describe and explain how the body adapts for hotter conditions.

the skin it takes energy to the environment and away from our body.

When we get cold the hairs on our skin stand up, trapping a layer of insulating air next to our body and decreasing the amount of energy lost to the environment.

We also shiver, which helps keep us warm as shivering increases respiration in the muscle cells which produces extra heat energy.

Finally, the capillaries in our skin can constrict or dilate which is called vasoconstriction or vasodilation and this helps us keep a constant temperature.

The answer describes vasoconstriction and vasodilation but does not go on to explain them so the student would lose a mark. To get full marks the answer would need to explain that, in vasodilation capillaries in the skin dilate, which increases blood supply to the skin and therefore increases the amount of heat transferred from the skin to the environment; in vasoconstriction, blood supply to the skin is reduced as capillaries get narrower, therefore there is less heat transferred to the environment.

Marks gained: [5 marks]

Measuring population size and species distribution

1.

Worked Example

Henry collects data to test his hypothesis: Creeping buttercups grow better in wetter places than bulbous buttercups.

Henry identifies an area where the ground changes from very wet to very dry. He uses a 20-metre transect line and a quadrat to estimate the percentage of each quadrat covered in each kind of buttercup, every 4 metres along the tape.

He also measures the percentage of water in the soil using a meter and a data logger. He does this twice in two different places. His results are shown below:

Quadrat	1	2	3	4	5
% water in the soil – test 1	23.2	34.5	44.3	56.6	68.8
% water in the soil – test 2	24.8	36.2	42.6	30.1	70.3
mean	_____	_____	_____	_____	_____
% cover of bulbous buttercups	60	50	30	20	0
% cover of creeping buttercups	0	20	30	50	70

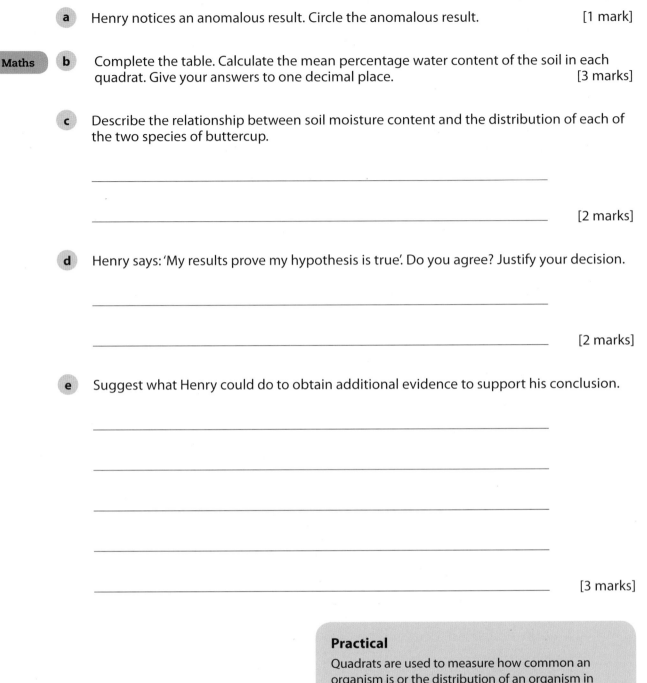

a   Henry notices an anomalous result. Circle the anomalous result.   [1 mark]

Maths   **b**   Complete the table. Calculate the mean percentage water content of the soil in each quadrat. Give your answers to one decimal place.   [3 marks]

c   Describe the relationship between soil moisture content and the distribution of each of the two species of buttercup.

_____

_____ [2 marks]

d   Henry says: 'My results prove my hypothesis is true'. Do you agree? Justify your decision.

_____

_____ [2 marks]

e   Suggest what Henry could do to obtain additional evidence to support his conclusion.

_____

_____

_____

_____

_____ [3 marks]

> **Practical**
>
> Quadrats are used to measure how common an organism is or the distribution of an organism in different areas, e.g. sunny and shady or trampled and un trampled. Transects are used to study how the distribution of an organism changes across an area, e.g. to find out if an organism becomes more common or less common as you move away from a footpath.

Cycling materials

1. Using words from the box, write a description of the water cycle.

| evaporate | condense | precipitation | transpiration |

_____

_____

_____

_____ [4 marks]

2. Explain why the water cycle is important to living organisms.

_____

_____

_____ [3 marks]

3. A scientist measured the concentration of carbon dioxide in the atmosphere every day for a year. She noticed the concentration varied very slightly according to the seasons as well as minor variations day to day.

Suggest **two** reasons for the slight variation in carbon dioxide concentration.

_____

_____

_____ [2 marks]

4. If all microorganisms in the world were suddenly destroyed, what effect would it have on the carbon cycle?

_____

_____

_____ [2 marks]

Remember

Practise interpreting carbon and water cycles presented in different ways. In the exam, you might be asked to identify, interpret or explain the key processes in carbon or water cycles that looks unfamiliar to you.

5. Describe, in detail, how carbon is continuously cycled through the ecosystem.

_____

_____

_____

_____

_____

_____ [6 marks]

Decomposition

1. The instruction manual for a garden compost bin recommends putting the bin in a sunny area and keeping the lid on. It also suggests turning the compost every month.

Explain why this advice is given.

_____

_____

_____

_____ [3 marks]

2. A class investigates the effect of temperature on the rate of decay of fresh milk by measuring pH change. Their method is shown below:

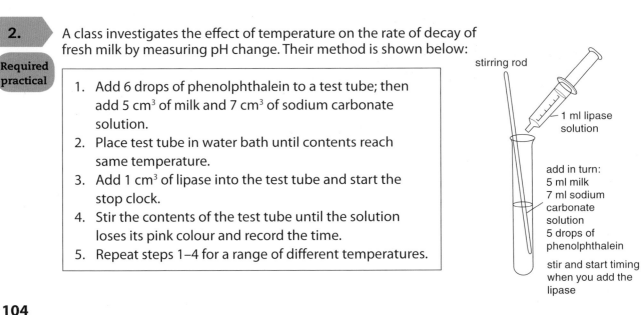

1. Add 6 drops of phenolphthalein to a test tube; then add 5 cm³ of milk and 7 cm³ of sodium carbonate solution.
2. Place test tube in water bath until contents reach same temperature.
3. Add 1 cm³ of lipase into the test tube and start the stop clock.
4. Stir the contents of the test tube until the solution loses its pink colour and record the time.
5. Repeat steps 1–4 for a range of different temperatures.

stirring rod

1 ml lipase solution

add in turn:
5 ml milk
7 ml sodium carbonate solution
5 drops of phenolphthalein

stir and start timing when you add the lipase

Worked Example **a** Explain why phenolphthalein changes from pink to colourless. [2 marks]

When the fat in the milk is broken down by the lipase, fatty acids are made. The fatty acids lower the pH of the mixture causing the colour change.

> This is a good, clear response, which would gain both marks.

Increasing temperatures from 0 °C to 45 °C reduces the time taken for the lipase to break down the fat in milk. However, over 45 °C, the time taken increases and in some cases the lipase does not work at all.

Marks gained: [2 marks]

b Explain why temperature affects the rate of decay of fresh milk in this way. [4 marks]

Because increasing the temperature increases the rate of reaction by increasing the collision rate between the enzyme and substrate molecules. This means the enzyme-controlled reactions that cause decay are increased and the rate of decay of milk increases up until 45 °C. Over 45 °C the

> This is a well-written answer that uses appropriate scientific terminology. This answer would gain full marks.

rate of enzyme reactions decreases as temperature increases until, at some point, the reaction stops altogether. This is because the active site of the enzyme is denatured by the high temperature and so the enzyme cannot continue to break down the fat.

Marks gained: [4 marks]

3. Explain why decay is vital for life.

_____ [1 mark]

4. The table shows how temperature affects the rate of biogas production in a generator in Egypt.

Maths

Temperature in °C	10	15	20	25	30	35	40
Volume of biogas produced each day in m³	0.40	0.45	1.40	1.60	2.90	3.35	3.15

Temperatures in Egypt may reach over 38 °C. Explain the advantage of the generator being mainly underground.

_____

_____ [2 marks]

Changing the environment

1. Describe and explain how deforestation can affect the distribution of species in an ecosystem.

_____

_____

_____ [3 marks]

2. The Aral Sea has shrunk since the 1960s after the rivers that fed it were diverted by irrigation projects. This caused environmental changes:

Higher Tier only

- an increase in salt concentration of the sea
- a decrease in oxygen concentration of the water
- higher sea temperatures.

Suggest and explain how these changes could have affected the distribution of plants and fish in the Aral Sea.

_____

_____

_____

_____ [3 marks]

> **Remember**
> This type of question requires you to apply your knowledge in a new context. You are **not** expected to have studied the Aral Sea. However, you need to apply your scientific knowledge and understanding of the impact of environmental changes. There will be more than one possible answer; and as long as you back up your suggestions with scientific explanations, you have a good chance of picking up marks.

3. Phytoplankton are tiny, plant-like organisms in the sea. They provide food for a wide variety of organisms.

Maths

Scientists worry that changes in global temperatures could impact on the distribution of phytoplankton in the ocean. They investigated how temperature affects the rate of photosynthesis of phytoplankton. The table shows their results.

Temperature (°C)	−1	3	7	11	15	19	23	27
Rate of photosynthesis (arbitrary units)	100	143	198	120	70	42	16	3

During the Antarctic spring, the mean temperature of the sea water is −0.8 °C. Use the data to suggest how global warming might affect the distribution of phytoplankton in the Antarctic Oc3ean.

_____

_____

_____

_____ [4 marks]

Effects of human activities

1. Define the term 'biodiversity'.

_____ [1 mark]

2. Explain why biodiversity is beneficial for an ecosystem.

_____

_____

_____ [3 marks]

3. Some coastal towns pour raw sewage through outfall pipes directly into the sea. Sewage is often high in phosphates and nitrates and contains bacteria such as *Salmonella* and *E. coli*. Suggest **two** impacts of releasing sewage into the sea.

_____

_____

_____ [2 marks]

4. Human population growth and increases in the standard of living in the UK have caused more landfill waste to be produced. If this waste is not handled correctly, it leads to pollution.

Evaluate the advantages and disadvantages of sending all our waste to landfill sites.

_____

_____

_____

_____

Command words

'Evaluate' means you should use the information given as well as your own knowledge and understanding to come to a judgement based on the evidence you give. Try to think of points for both sides of the argument, e.g. advantages as well as disadvantages.

_____

_____

_____ [5 marks]

Global warming

1. The gases that cause global warming occur naturally in the atmosphere, but human activities have increased the level of these gases.

List **two** human activities that have increased the level of these gases.

_____

_____ [2 marks]

2. Give **two** reasons why the concentration of methane is the atmosphere has increased over recent years.

_____

_____ [2 marks]

3. The graph shows the temperature difference (compared to 1880 global average temperature) between 1880 and 2012.

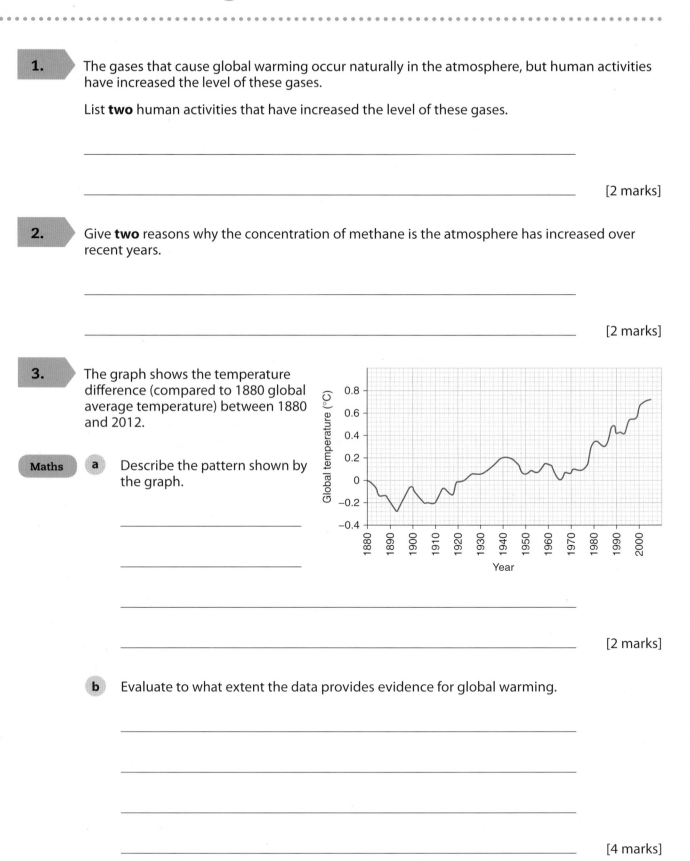

Maths **a** Describe the pattern shown by the graph.

_____

_____

_____

_____ [2 marks]

b Evaluate to what extent the data provides evidence for global warming.

_____

_____

_____

_____ [4 marks]

Maintaining biodiversity

1. Monoculture is the practice of growing one type of crop, usually on a large area of land.

Programmes have been set up to encourage farmers to:

* reintroduce hedgerows
* leave a margin around field edges for wild flowers and grass.

Explain how these measures could reduce the impact of monoculture on the ecosystem.

_____

_____

_____ [3 marks]

2. The African elephant is classed as a threatened species due to loss of habitat, poaching and being hunted for meat.

Over the past few decades many conservation programmes have been introduced to increase elephant numbers.

Using the information given and your own knowledge and understanding, describe and explain the conflicting pressures between local residents and conservation programmes. [4 marks]

The natural habitat for some African elephants is turned into cropland by locals due to an increasing population and demand to produce more food from crops to feed local populations and make money.

Locals may see this as necessary to survive whereas conservation programmes want to stop destruction of the elephant habitat and want to make sure the elephants' habitat and migratory routes are not broken up.

Some locals poach elephants for their tusks, which can sell for a lot of money. The meat from elephants can also sell for a good price. Locals, that might have no other source of income, can earn a lot of money from poaching.

Conservation groups put effort into stopping poaching and educating people about the cruelty involved and the need to protect the elephant species. However, locals might not

> This is a detailed and well-structured answer that would gain full marks. The answer uses the information given at the start of the question. If a question states 'using the information given' make sure you do use some of the information in your answer as it is likely to be in the mark scheme.
>
> The answer shows a good understanding of the conflicting pressures.

understand the reasons to protect a species and the need to earn money and bring food to the table is more important to them than protecting the elephant.

Marks gained: [4 marks]

3. Explain the potential ecological benefits of protecting the Amazon rainforest.

_____

_____

_____

_____

_____

_____

> **Remember**
> You need to be able to describe both negative and positive human interactions in an ecosystem and explain their impact on biodiversity. Make sure you know examples from a range of ecosystems. Questions on this topic could feature unfamiliar ecosystems like mangroves and coral reefs so you will need to apply your knowledge and understanding.

_____

_____ [6 marks]

Biomass in an ecosystem

1. Trophic levels are the different stages of a food chain. Describe the difference between the types of organisms you would find in trophic levels 1 and 2 in an ecosystem.

_____

_____

_____ [4 marks]

2. The data in the table shows the biomass found at each trophic level for an English woodland.

Maths

Trophic level	g/m^2
1	6000
2	500
3	50

Use the information in the table to construct a pyramid of biomass for this food chain.

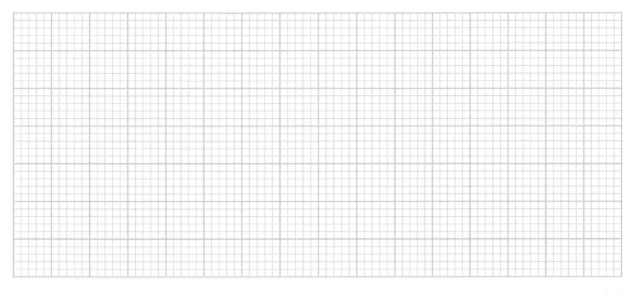

[2 marks]

3. Pyramids of biomass for a marine ecosystem may sometimes show producers such as phytoplankton with a smaller biomass than primary consumers such as zooplankton. Suggest a reason for this.

_____

_____ [1 mark]

4. The table shows the amount of energy transferred at each trophic level in a lake.

Trophic level	1	2	3	4
Name of organism	waterweeds	tadpoles	minnow	pike
Energy transferred to next level (kJ/m²/year)	80 000	7900	760	63

Maths **a** The energy transfer between trophic levels 1 and 2 is only 9.9%. Calculate the efficiency of energy transfer between trophic levels 2 and 3 and trophic levels 3 and 4. Give your answers to one decimal place.

_____

_____

Maths

To calculate energy efficiency, you need to use this formula:

Efficiency =

$$\frac{\text{energy available after the transfer}}{\text{energy available before the transfer}} \times 100$$

It is a good idea to look at the numbers and work out roughly what sort of percentage you would expect.

Energy transfer between trophic levels 2 and 3 = _____ % [1 mark]

Energy transfer between trophic levels 3 and 4 = _____ % [1 mark]

b The efficiency of energy transfer between different levels is very low. On average, only about 10% of the biomass is passed onto the next level. Explain why.

_____

_____

_____

_____ [4 marks]

Food security

1. Describe what is meant by the term 'food security'.

_____ [1 mark]

2. List **four** biological factors that are threatening food security.

_____

_____

_____

_____ [4 marks]

3. Describe and explain how fishing quotas could conserve fish stocks.

_____

_____

_____ [2 marks]

4.

Maths

Scientists investigated the amount of energy needed to produce 1 kg of different food products.

Their results are shown below:

Food product	Amount of energy (kW hours) needed to produce 1 kg food product
Corn	0.5
Wheat	0.7
Chicken	4.2
Pork	12.5
Beef	32.6

Use the information in the table to suggest what effect decreasing consumption of meat products could have on global food security.

_____

_____

_____ [3 marks]

Role of biotechnology

....

1. The fungus *fusarium* is used to produce which food product?

_____ [1 mark]

2. Give **two** advantages and **two** disadvantages of using genetically modified crops.

Advantages: _____

_____

Disadvantages: _____

_____ [4 marks]

3. For many years, insulin was obtained by purifying it from the pancreas of cows and pigs slaughtered for food. This was time-consuming and expensive. It also raised ethical issues for people with diabetes who could not use pig's insulin because of religious beliefs or vegetarianism.

Insulin is now made by genetically engineered microbes. The genetically modified bacteria are grown in large fermentation vats containing all the nutrients needed for growth. This allows human insulin to be produced in large quantities and then purified. This form of insulin is absorbed more rapidly than animal-derived insulin and acts more quickly.

Evaluate the **advantages** and **disadvantages** of using of biotechnology to produce insulin.

> **Command words**
>
> Remember that the command word 'evaluate' means you should use the information supplied, as well as your own knowledge and understanding, to write about the advantages and disadvantages of using biotechnology. At the end, you should say whether or not it is a good idea, based on the points you have raised.

[6 marks]

Plant and animal cells (eukaryotic cells)

1. **a** A [1 mark]

   b ribosome [1 mark]

   c D [1 mark]

2. **a** Cell walls are supporting structures that help the plant stay turgid/have a fixed shape/stay upright and protect it from injury. [1 mark] Animal cells do not need cell walls as they have the skeletal system to protect the organs and cushion them against any external injuries. [1 mark]

   b Chloroplasts; [1 mark] a permanent vacuole filled with cell sap. [1 mark]

3. Start with the lowest power objective on the microscope. Insert graticule into eyepiece and stage micrometer onto stage of microscope. Align the scale on the stage micrometer with the scale of the eyepiece graticule. Use the reading taken from the scales to calculate the calibration factor for the objective lens in use. [4 marks]

Bacterial cells (prokaryotic cells)

1. A – plasmid; B – flagellum; C – single loop of DNA. [3 marks]

2. Bacteria; archaea. [2 marks]

3. Cytoplasm – eukaryotes and prokaryotes; [1 mark] cell wall – prokaryotes; [1 mark] genetic material enclosed in a nucleus – eukaryotes; [1 mark] genetic material in single DNA loop – prokaryotes. [1 mark]

4. Worked example - full answer given in workbook.

Size of cells and cell parts

1. A: 0.00002; [1 mark] B: 0.12; [1 mark] C: 150. [1 mark]

2. 28 μm into metres = 0.000028 m; [1 mark] which is 2.8×10^{-5} m (in standard form). [1 mark]

3. Magnification = $30 \times 10 = \times300$; [1 mark] real size = $\frac{9}{300} = 0.03$ mm; [1 mark] which is ($\times 1000$) 30 μm. [1 mark]

Electron microscopes

1. Magnification is how much bigger an image appears; [1 mark] resolution refers to the resolving power, or the ability to distinguish between two points, meaning a higher resolution gives a sharper image. [1 mark]

2. It allows living cells to be viewed in nano-scale detail so living cancer cells can be observed. This could improve our understanding of the mechanism of cancer and its treatment. [2 marks]

3.

Level 3: The response compares most of the advantages and disadvantages of both the electron microscope and the light microscope.	5–6
Level 2: The response compares some of the advantages and disadvantages of the electron microscope and the light microscope.	3–4
Level 1: The response compares at least one advantage and one disadvantage of the electron microscope with the light microscope.	1–2

Indicative content

- You can obtain better quality images of the internal structure of the cells using an electron microscope rather than a light microscope.
- Electron microscopes have a better resolution than light microscopes.
- Electron microscopes have higher magnification than light microscopes.
- Electron microscopes are very expensive compared with light microscopes.
- Electron microscopes are much bigger and are not portable like light microscopes.
- Specialist training is needed to use electron microscopes whereas only basic training is needed to use a light microscope.
- The specimen must be dead for viewing with an electron microscope whereas specimens don't necessarily have to be dead for viewing with a light microscope.
- Specimen has to be prepared in a vacuum for an electron microscope, but not for a light microscope.
- It takes much longer to prepare and study a specimen for viewing in an electron compared with a light microscope.

Growing microorganisms

1. 128 cells means 7 divisions must have taken place ($2^7=128$); [1 mark] $\frac{280}{7} = 40$ minutes. [1 mark]

2. **a** Divide diameter of zone B by 2 to get radius: $\frac{12}{2} = 6$ mm. [1 mark] Use πr^2 to calculate the area of B: $3.14 \times 36 = 113$ mm^2. [1 mark]

   b That A was a more effective antibiotic as the zone of inhibition was more than that of antibiotic B. [1 mark]

c Any **three** from: sterilising the petri dishes before use by heating to a high temperature to kill microorganisms; using a sterilised inoculation loop to transfer bacteria to the culture medium by passing it through a hot flame; after transferring bacteria onto agar jelly, lightly tape lid of petri dish on to stop microorganisms in the air getting in; store petri dish upside down to stop drops of condensation falling onto the agar surface. [3 marks]

Cell specialisation and differentiation

1. Cell differentiation allows a cell to become specialised for its function; [1 mark] and therefore allows organisms to function more efficiently/improves chances of survival. [1 mark]

2. Both sperm and egg cells are haploid so the fertilised egg will be diploid. The cytoplasm of the egg cell is packed with nutrients to supply the fertilised egg with energy and raw materials for growth and development. [2 marks]

3. More ribosomes; [1 mark] because ribosomes are the cell organelles that make proteins. [1 mark] The main function of the acinar cells is to produce enzymes, which are made from proteins, so you'd expect more ribosomes in these cells. [1 mark]

4. Most animal cells lose the ability to differentiate after they become specialised. Most plant cells do not lose the ability to differentiate. Cells that differentiate in mature animal cells are usually for repairing or replacing cells. [2 marks]

Cell division by mitosis

1. Chromosomes are long lengths of DNA. [1 mark] Short sections of DNA form genes. [1 mark]

2. The cytoplasm divides into two and the new cell membrane forms two new cells – 5. [1 mark] Further growth occurs and the DNA is checked for errors – 3. [1 mark] Mitosis occurs and the chromosomes move apart, forming two new nuclei – 4. [1 mark] The DNA replicates – 2. [1 mark]

3. Cells spend $14/170 \times 100 = 8.235$ % of time in last stage of mitosis [1 mark] $8.235/100 \times 23 = 1.8941$ hours [1 mark] which is 1.89 hours to 2 significant figures [1 mark]

4. Worked example - full answer given in workbook.

Stem cells

1. Unspecialised cells that can differentiate into many types of cells. [1 mark] They are important for growth and repair. [1 mark]

2. Stem cells can be grown in a laboratory to form clones; one stem cell can produce more stem cells. [2 marks]

3. It is a quick and economical method of producing large numbers of identical crop plants that have desirable features such as disease resistance. [1 mark] Rare species can be cloned to protect them from extinction. [1 mark]

4. Use the meristem cells of Dutch elm tree that is naturally resistant to Dutch elm disease; to create and grow clones. [2 marks]

5.

Level 3: The response considers most of the advantages and disadvantages of the use of embryonic stem cells in medicine. An overall judgement is provided based on the points discussed.	5–6
Level 2: The response considers some of the advantages and disadvantages of the use of embryonic stem cells in medicine. An overall judgement is attempted based on the points discussed.	3–4
Level 1: The response considers at least one advantage and disadvantage of the use of embryonic stem cells in medicine. An overall judgement may not be attempted.	1–2

Indicative content

Advantages:
- Embryonic stem cells have potential to differentiate into many types of cell.
- Embryonic stem cells are easier to grow in culture.
- Embryonic stem cells are not rejected by patients.
- Has potential to treat many illnesses and conditions.
- Has potential to lessen suffering and improve lives of those with illnesses, if treatment is successful.

Disadvantages:
- Stem cells could be contaminated with a virus which could be passed onto the patient.
- Requires donated eggs or eggs to be made artificially.
- Some people may disagree with use of stem cells on religious grounds.
- Expensive and time-consuming.
- Risk that stem cells may divide out of control and cause cancer.
- There are many moral and ethical issues raised;
  ○ some people may feel that human embryos should not be used in this way as each one is a potential life;
  ○ there are other sources of stem cells that scientists could use that are more ethical, e.g. unwanted embryos from fertility clinics.

Diffusion in and out of cells

1. The movement of molecules from an area of higher concentration to an area of lower concentration. [1 mark]

2. If the air is warmer, the molecules of carbon dioxide have more energy and move faster so the rate of diffusion is faster. [2 marks]

3. The concentration of oxygen outside of the muscle cell will be higher than inside; because the muscle cell uses oxygen during respiration to provide energy for contraction; therefore, the oxygen diffuses down the concentration gradient into the muscle cell. [3 marks]

4. a A: Arrow showing the glucose diffusing out of the cell. [1 mark] Explanation: glucose will diffuse out of the cell because there is a higher concentration inside compared to outside. [1 mark]

   B: No arrow needed as no net movement. [1 mark] Explanation: there will be no net movement of glucose because the concentration of glucose is the same inside and outside the cell. [1 mark]

   C: Arrow showing glucose entering the cell. [1 mark] Explanation: the glucose will diffuse into the cell because there is a higher concentration outside compared to inside. [1 mark]

   b C; [1 mark] because the concentration gradient is highest between the inside and outside of the cell. [1 mark]

5. Worked example - full answer given in workbook.

Exchange surfaces in animals

1. Hippopotamus. [1 mark]

2. Any **three** from: glucose; oxygen; carbon dioxide; water; urea. [3 marks]

3. a Surface area = 54 cm²; volume = 27 cm³; SA:V = 2:1. [3 marks]

   b A, because it has the highest surface area to volume ratio. [1 mark]

4. Their round shape decreases their surface area to volume ratio so less heat is lost. This increases the chance of survival as they cannot rely on their metabolic rate to keep warm. [2 marks]

5. Worked example - full answer given in workbook.

Osmosis

1. The diffusion of water from a dilute solution to a concentrated solution through a partially permeable membrane. [2 marks]

2. a Because the potato chips might have had slightly different masses at the start. [1 mark]

   b Any **two** from: volume of the sugar solution; type of sugar used; type of potato/same potato; temperature of the solution.

c Both axes correctly labelled; both axes appropriately scaled; points plotted correctly; correct line of best fit. [4 marks]

d In the range of 0.42–0.44 M. [1 mark]

e The potato chip in tube 2 gained mass because the concentration of the sugar solution outside the potato was less than that of the fluid inside the potato. Therefore, water particles moved into the potato by osmosis, causing it to gain mass. The potato chip in tube 7 lost mass because the concentration of the sugar solution was higher than that of the fluid inside the potato so water particles moved out of the potato and into the solution by osmosis, causing the potato to lose mass. [4 marks]

Active transport

1. Worked example - full answer given in workbook.

2. The concentration of mineral ions is higher in the plant than in the soil, so the ions need to be moved against the concentration gradient. This cannot occur with diffusion as this only takes place down a concentration gradient. The minerals cannot be absorbed by osmosis as this is the movement of water only. [3 marks]

3. Mitochondria are the site of respiration. The more mitochondria, the more respiration and therefore the more energy available for active transport. Active transport is needed to absorb minerals from the soil and prevent mineral deficiencies. [3 marks]

4. There are times when there is a lower concentration of glucose and amino acids in the gut than in the blood. During these times, the concentration gradient is the wrong way for diffusion, so active transport is required. Therefore, without active transport we wouldn't be able to absorb vital substances such as glucose and we would starve. [3 marks]

Section 2: Organisation

Digestive system

1. A group of organs working together to perform a particular function. [1 mark]

2. To build new carbohydrates, lipids and proteins; some glucose is used for respiration. [2 marks]

3. The small intestine is where food is digested and soluble food molecules are absorbed; the large intestine is where water is absorbed back into the blood from undigested food, leaving faeces. [2 marks]

4. 9 m = 9 000 000 μm, which is 9.0×10^6. [1 mark]

5. Bile emulsifies fats into smaller droplets to increase the surface area. Bile is alkaline and this, together with the

large surface area, increases the rate of fat digestion into fatty acids and glycerol. [3 marks]

6. They provide a large surface area so there is more area for diffusion to take place over. They have a good blood supply with a network of capillaries inside each villus, which assists in quick absorption. They have a single layer of surface cells so the distance the nutrients have to diffuse across is small which increases the rate of diffusion. [3 marks]

Digestive enzymes

1. Proteases – break down proteins into amino acids. [1 mark]

    Lipases – break down lipids into fatty acids and glycerol. [1 mark]

2. Put some of the food sample into a test tube. Add 3 drops of Sudan III stain solution and gently shake the tube. If the cake contains lipids (fats) the mixture will separate into two layers, the top will be bright red. [3 marks]

3. D, B, C, A. [1 mark]

4. It is a non-reducing sugar; Benedict's only tests for reducing sugars. [2 marks]

Factors affecting enzymes

1. The theory explains why enzymes are specific because both the substrate and the enzyme have a unique shape like a lock and key. The key represents the substrate and the lock represents the enzyme's active site. Only the substrate with the specific shape that fits exactly into the active site of the enzyme will work/form an enzyme–substrate complex. [3 marks]

2. a   The temperature of the water in the beaker is difficult to control with a Bunsen burner. [1 mark]

    b   Use a water bath. [1 mark]

    c   A: As temperature increases, particles move faster; [1 mark] so there is a greater chance of a substrate molecule entering an active site and being changed. [1 mark]

    B: The optimum temperature is reached. [1 mark] The enzyme is working fastest because many fast-moving substrate molecules enter and fit easily into the active site. [1 mark]

    C: When temperature is above the optimum, the active site changes shape/the enzyme denatures so the substrate molecule no longer fits; [1 mark] and the rate of reaction decreases until it eventually stops. [1 mark]

The heart and blood vessels

1. Any **two** from: blood pressure is higher, especially to the body; there is higher blood flow to the body tissues; oxygenated blood is separated from deoxygenated blood. [2 marks]

2. $\frac{1757}{3.5} = 502$ ml/min. [1 mark]

3. Implant a pacemaker. This would produce an electric current to keep the heart beating regularly. [2 marks]

4. Worked example - full answer given in text.

Blood

1. It transports substances around the body. [1 mark]

2. 3737.8 cm^3. [1 mark]

3. a   They help blood clot at a wound to prevent excessive blood loss and microorganisms from entering the wound. [2 marks]

    b   Excessive bruising. [1 mark] Excessive bleeding. [1 mark]

4. Biconcave-shaped; giving a large surface area to volume ratio; increases efficiency of diffusion of oxygen in and out of the cell/reduces diffusion distance to centre of cell; no nucleus, increasing space available for haemoglobin, contains haemoglobin, which carries oxygen. [5 marks]

Heart–lungs system

1. Pulmonary artery. [1 mark] It is deoxygenated because it is carrying blood that has been all around the body back to the lungs [1 mark] so most of the oxygen will have diffused into body cells. [1 mark]

2. Fabio's breathing rate = $\frac{108}{12} = 9$ breaths per minute;

    Idris's breathing rate = $\frac{91}{7} = 13$ breaths per minute.

    Therefore, Idris has the faster breathing rate. [2 marks]

3. If the phrenic nerve is damaged, it would not send an electrical message to the diaphragm to tell it to contract. The diaphragm changes the volume and pressure of the thoracic cavity, causing air to move in and out so gas exchange can take place. Therefore, Evie could die due to lack of oxygen. [3 marks]

4. Blood could pass from the left atrium to the right atrium; so oxygenated blood would mix with deoxygenated blood. Some oxygenated blood would get pumped to lungs and the amount of oxygen dissolved in blood pumped to the rest of the body could be slightly lower than if there was no hole. [3 marks]

Coronary heart disease

1. A stent is a wire mesh tube that can be inserted into arteries to widen them/keep them open. It keeps the coronary artery open, ensuring blood can pass through to the heart muscle so the person's heart can keep beating. [2 marks]

2. Any **two** from: risk of complications during surgery; risk of infection after surgery; fatty deposits can rebuild; risk of thrombosis/blood clot forming near stent. [2 marks]

3. Statins stop the liver producing as much LDL-cholesterol and increase the amount of HDL-cholesterol. HDLs further reduce amount of LDL-cholesterol in blood. Less LDL-cholesterol means slower rate of fatty deposits forming in coronary arteries. [3 marks]

4. Benefit to Betty – it could extend her life and improve her quality of life. [1 mark] Risks to Betty – her age, high blood pressure and frailness means it could be risky to undergo the major surgery required to replace the valve because she might not be able to fight infection or survive other complications associated with surgery like a younger, healthier patient. [1 mark] Increased risk of blood clots so she might have to take anticoagulants, which also carry risks. [1 mark] Final mark for weighing up risks and benefits to come to a conclusion. [1 mark]

5. Worked example - full answer given in workbook.

Risk factors for non-infectious diseases

1. An event or circumstance that is linked to an increase in the chance a person will develop a certain disease or condition during their lifetime. [1 mark]

2. People from deprived areas are more likely to (accept any **two** from): smoke; have a poor diet; not exercise. [2 marks]

3. Time lag involved between risk factor and onset of non-communicable disease; many factors involved make it difficult to prove a causal mechanism; some risk factors are not capable of directly causing a disease, but are related to another risk factor, e.g. lack of exercise. [3 marks]

4. This is only a general trend and there are discrepancies in the data shown. E.g. the coronary heart disease incidence is higher for those with a low BMI than those with a medium BMI **or** there was no difference between the coronary heart disease incidence for those with a high or highest BMI. The conclusion is not fully supported by all the data. [3 marks]

Cancer

1. A benign tumour is slow-growing, a malignant tumour is faster; a benign tumour often has a clear border, a malignant does not necessarily; a benign tumour is not cancerous, a malignant tumour is; a benign tumour rarely spreads, a malignant tumour spreads to other body tissues easily. [4 marks]

2. Worked example - full answer given in workbook.

Leaves as a plant organ

1. A: Vascular bundle/phloem and xylem. [1 mark] B: Upper epidermis. [1 mark] C: Spongy mesophyll layer. [1 mark]

2. Xylem vessels deliver water and minerals to the leaf so damage to these vessels would mean the leaf would not receive enough water. The leaf would not be able to photosynthesise without water and could therefore not produce glucose for growth and respiration for energy. Minerals normally carried in xylem would not be effectively transported to plant tissue so symptoms of mineral deficiencies might appear. Phloem vessels would not be able to translocate glucose effectively, which would reduce the amount available for growth and respiration so the plant would not grow well and would probably die. [5 marks]

3. The lower epidermis contains stomata which let carbon dioxide diffuse into the leaf. Guard cells can open and close stomata to regulate the exchange of gases. The spongy mesophyll contains air spaces to increase the surface area, which increases the rate of diffusion. [3 marks]

Transpiration

1. The transpirational pull: when the plant loses water through transpiration from the leaves, water from the stem and roots moves upwards into the leaves. Cooling of the plant: the loss of water vapour from the plant cools down the plant when the weather is very hot. Plant structure: young plants or plants without woody stems require water for structural support. Transpiration helps maintain the turgidity in plants. [3 marks]

2. a Normal conditions – 2.3/25 = 0.092; decreased air temperature – 2.9/25 = 0.116; increased humidity – 3.4/25 = 0.136. [1 mark]

   b At increased air temperatures the rate of transpiration is higher because water molecules move faster and the rate of evaporation from stomata is therefore much faster. When the humidity is low the concentration gradient between the inside of the leaf stomata and the atmosphere is high and the rate of diffusion will be faster so rate of transpiration will also be increased. [4 marks]

Translocation

1. Phloem tubes are living cells joined end to end with small pores; phloem transports food substances in both directions. [2 marks]

2. Worked example - full answer given in workbook.

3.

Level 3: The response gives a clear and detailed description and explanation of transpiration and translocation.	5–6
Level 2: The response gives a reasonable description and explanation of transpiration and translocation.	3–4
Level 1: The response gives basic information about transpiration and translocation.	1–2

Indicative content	
Transpiration: • is loss of water from a plant. • explains how water moves up the plant against gravity. • takes place in tubes made of dead xylem cells. • Water on the surface of spongy and palisade cells (inside the leaf) evaporates and diffuses out of the leaf so more water is drawn out of the xylem cells inside the leaf to replace it. • This means more water is drawn up from the roots; • producing a constant stream of water and minerals through the plant. **Translocation:** • is the movement of food substances from leaves to other tissues throughout the plant. • The sugars made in leaves by photosynthesis may be needed by other parts of the plant; • so food substances/sugar are translocated from leaves to where the carbohydrate is needed, e.g. growing tissue or storage tissue. • happens in phloem tubes • Translocation of substances in the phloem can be in any direction.	

Section 3: Infection and response

Microorganisms and disease

1. **a** Diseases caused by pathogens/examples of pathogens; that are spread by direct contact/water/air; any **one** example from measles; mumps; rubella; colds; flu/impetigo; any other infectious disease. [3 marks]

   b Produce poisons/toxins/damage cells/tissues. [1 mark]

2. Gonorrhoea: treated with antibiotics; malaria - caused by a protist; measles - can reduce spread by using handkerchiefs. [3 marks]

3.

Level 3: The response includes a number of key points and at least one comparison.	5–6
Level 2: The response includes a number of key points.	3–4
Level 1: The response includes one or more key features of spread of disease.	1–2

Indicative content:	
• (both) reproduce rapidly in the body; • bacteria produce toxins; • (bacterial) toxins damage tissues; • bacteria do not enter cells whereas viruses live/reproduce inside cells, where they cause cell damage; • bacteria considered living whereas viruses are non-living; • bacteria killed by antibiotics.	

Viral diseases

1. **a** Fever/skin rash. [1 mark]

   b Inhalation of droplets (from sneezes/coughs). [1 mark]

   c Fatal; if complications arise/description of complications (e.g. hepatitis, meningitis). [2 marks]

2. **a** Number of deaths (in 2016) is 100% – 84% = 16% (of deaths in 2000); 550,100 x 16/100 = 88,016 [2 marks]

   b Less chance of contact with someone who has the disease/herd immunity. [1 mark]

3. **a** AIDS. [1 mark]

   b Immune cells/T-lymphocytes. [1 mark]

   c Antiretroviral (drugs). [1 mark]

Bacterial diseases

1. No; different bacteria have different shapes/sizes; credit examples. [2 marks]

2. **a** Raw eggs; partially cooked meat; unpasteurised milk (products). [2 marks]

   b Sexual contact. [1 mark]

   c Antibiotics; using barrier methods of contraception/condoms. [2 marks]

3. The pathogen enters an organism; the pathogen reproduces rapidly in ideal conditions to increase numbers/incubation period; pathogens make harmful toxins, which build up; symptoms develop, for example, fever and a headache. Accept answer with named bacterial infection and its stages. [4 marks]

Malaria

1. **a** Parasite. [1 mark]

   b Stages 1 **and** 3. [1 mark]

   c Vectors. [1 mark]

2. Prevent mosquitoes from breeding; kill larvae/use chemicals/introduce sterile males/destroy breeding grounds/empty, drain or cover all things that hold water; prevent mosquitoes biting; use chemicals/ sprays/repellents/nets. [4 marks]

Answers

Human defence systems

1. a Kills the majority of pathogens that enter via the mouth/food. [1 mark]

   b Physical barrier/keratinisation; (sebaceous) fluids contain chemicals/antimicrobials. [2 marks]

   c Cilia; mucus secretion. [2 marks]

2. Platelets in the wound exposed to air at site of wound; (exposure to air) makes platelets produce protein fibres/mesh; platelets and red blood cells are caught in fibres to form a clot; wound sealed to prevent pathogens from entering. [4 marks]

3. Reproducing rapidly; producing poisons/toxins; (that) damage tissues. [3 marks]

4. Any **two** from: broken down; by enzymes/proteases; into amino acids. [2 marks]

5. Phagocytosis; engulfing and ingesting pathogens; antibody production; identify and neutralise pathogens; antitoxin production; neutralise toxins/kill pathogens. [6 marks]

Vaccination

1. a Viruses live inside cells; (so) difficult for drugs/immune system to reach. [2 marks]

2. a Inactive/dead pathogen. [1 mark]

   b Vaccination results in less chance of coming into contact with the disease/herd immunity; idea that (more) infectious diseases are more likely to be passed on/spread. [2 marks]

3. Any **three**, in a logical sequence, from: (vaccination) introduces antigens/harmless pathogens; targets immune response; white blood cells produce antibodies; triggers production of memory lymphocytes. [3 marks]

4. Any **three** sequential points from: after A, there are no memory lymphocytes so antibody production is slower and fewer antibodies are produced; at B, memory lymphocytes are already present so antibody production is immediate; at A, a low level of antibodies produced/time delay to trigger immune response; at B, memory lymphocytes triggered; so at B, higher level of antibody production/antibodies produced more quickly. [3 marks]

Antibiotics and painkillers

1. Infectious disease are caused by pathogens; painkillers do not kill pathogens. [2 marks]

2. a No, some volunteers should be given no drugs/placebo; for comparison/to see if **A** and **B** really work. [2 marks]

   b Any **two** from: age; gender; ethnicity; severity of pain/how long they had pain before trial; type of pain/illness/site of pain; weight/height; other medical issues. [2 marks]

3. Antibiotics will kill bacteria but not viruses; gonorrhoea is caused by bacteria and HIV by a virus. [2 marks]

4. Overuse/misuse of antibiotics; bacteria mutate; antibiotics kill non-resistant strains; resistant bacteria reproduce. [4 marks]

Making and testing new drugs

1. Scientists running the trial but not those directly involved with the patients/volunteers. [1 mark]

2. a 110. [1 mark]

   b 4. [1 mark]

   c Calculate the percentage developing heart disease (with each drug). [1 mark]

3.

Level 3: A clear, detailed and logical description of all stages of testing.	5–6
Level 2: A partial but logical description of most stages of testing.	3–4
Level 1: Some relevant points made about the stages of testing.	1–2

   Indicative content:

   - Testing for toxicity/efficacy/dose.
   - Pre-clinical testing on cells/tissues/live animals.
   - Clinical trials on healthy volunteers/patients.
   - Start with very low doses.
   - To check for safety/side effects.
   - Further trials to find the optimum dose.
   - Double blind trial.
   - Some volunteers/patients given a placebo.
   - Random allocation of patients to groups.
   - So no one knows who has placebo/the new drug.
   - Peer review of results.

Monoclonal antibodies

1. a To stimulate production of lymphocytes (which produce antibodies). [1 mark]

   b So that hybridoma cells will divide quickly (like tumour cells); (and) produce antibodies (like lymphocytes). [2 marks]

2. a An antibody that is of just one type/all same type. [1 mark]

   b (Monoclonal antibodies) bind tightly to specific molecules on cell surface; therefore, monoclonal antibodies detect presence of gonorrhoea bacteria because there are different molecules on surface of chlamydia. [2 marks]

   c Create more side effects than expected/may provoke immune response in patients. [1 mark]

3. Radioactive substances/toxic drugs/chemicals to stop cells growing and dividing; delivered to cancer cells (only); without harming other cells in the body. [3 marks]

Plant diseases

1. a (Rose) black spot. [1 mark]

   b Fungal spores in water/wind. [1 mark]

   c Fungicides; removing/destroying affected leaves. [2 marks]

2. Worked example - full answer given in workbook.

3. Any **two** from: suck sap/remove sugars/nutrients; reduce sugar/energy available for growth; aphids carry viruses/spread disease. [2 marks]

Identification of plant diseases

1. a Nitrate makes proteins/nucleic acids; poor/stunted growth. [2 marks]

   b Any **three** from: the movement of chemicals; across a permeable membrane; against concentration gradient; requires energy. [3 marks]

   c Magnesium. [1 mark]

   d Yellow leaves/chlorosis (do not accept just poor growth). [1 mark]

2. Reference to gardening manual/website; taking infected plants to a laboratory to identify the pathogen; using testing kits with monoclonal antibodies. [3 marks]

3. Any **four** from: (stunted) growth; spots on leaves; areas of decay; growths; appearance of leaves/other plant parts; discolouration; presence of pests. [4 marks]

Plant defence responses

1. Layers of dead cells around stems to prevent pests from damaging living cells underneath and introducing pathogens; waxy leaf cuticle to prevent pathogens entering the epidermis; cellulose cell walls to prevent pathogens entering cells. [3 marks]

2. Chemical defence; prevents bacteria from attacking/infecting. [2 marks]

3. Tricks butterflies into not laying eggs; to prevent the caterpillars eating/damaging the plant. [2 marks]

4. Stinging hairs; stops animals eating; prevents damage to plant. [3 marks]

Section 4: Photosynthesis and respiration reactions

Photosynthesis reaction

1. Energy is transferred from the environment to the chloroplasts by light. [1 mark]

2. carbon dioxide + water $\xrightarrow{\text{light}}$ glucose + oxygen [2 marks]

$$6CO_2 + 6H_2O = C_6H_{12}O_6 + 6O_2 \quad \text{[2 marks]}$$

3. A = 12 cm^3; B = 11 cm^3. [2 marks]

4. Deforestation would result in less photosynthesis; less photosynthesis would cause less carbon dioxide to be absorbed from the atmosphere; carbon dioxide concentration would increase; less oxygen would be released from photosynthesis; oxygen concentration would decrease. [4 marks]

Rate of photosynthesis

1. a

Level 3: The response gives a clear and detailed description of a method with apparatus that would produce valid results, including reference to how variables would be controlled.	5–6
Level 2: A method involving pondweed and varying light intensity is given, but some parts may lack detail. Answer includes at least one control variable.	3–4
Level 1: The response includes simple statements relating to relevant apparatus or the method involving pondweed and light, but answer lacks detail and structure.	1–2
Indicative content: • Description of how the apparatus would be used. • Use of ruler to measure distance of light from pondweed. • Reference to varying distance of light from pondweed. • Accept alternative methods to alter light intensity. • Measure number and volume of gas produced using a capillary tube and syringe. • Same length of time. • Reference to control of temperature. • Reference to control of carbon dioxide in water. • Do repeats and calculate a mean.	

   b Freshly cut pondweed is more likely to produce gas bubbles. Adding sodium hydrogen carbonate indicator increases the carbon dioxide concentration of the water, which increases rate of photosynthesis. [2 marks]

2. Worked example - full answer given in workbook.

3. Worked example - full answer given in workbook.

4. a 0.01 [1 mark]

   b 0.0025 [1 mark]

5. Use a light meter. [1 mark]

Answers

Limiting factors

1. a Both axes and lines correctly labelled; both axes appropriately scaled; points plotted correctly; curved line of best fit. [4 marks]

b At the start, both lines show that, as the light intensity increases, the rate of photosynthesis steadily increases. At the low carbon dioxide concentration, the rate of photosynthesis begins to level off at a light intensity of about 0.46 AU. This is the point at which the low carbon dioxide concentration becomes the limiting factor. At the higher carbon dioxide concentration, the rate of photosynthesis levels off at a higher light intensity of about 0.90 AU. At this point the carbon dioxide concentration becomes the limiting factor again. [5 marks]

The uses of glucose from photosynthesis

1. Plant cells respire all the time, but some also carry out photosynthesis when light is available; only some plant cells can carry out photosynthesis, some – such as root hair cells – do not. [2 marks]

2. Some is converted into lipids (fats and oils) – these are stored in the seeds; some is converted into starch. [2 marks]

3. Higher concentration of glucose inside cell would cause movement of water into the cell by osmosis. This could cause cell to swell and eventually burst. [2 marks]

4. Field C because it has the highest nitrate percentage. Nitrates combined with glucose form amino acids, which make proteins. Proteins are essential for healthy growth. [3 marks]

Cell respiration

1. An exothermic reaction that transfers energy to the environment; a series of reactions catalysed by enzymes. [2 marks]

2. glucose + oxygen $\longrightarrow$ carbon dioxide + water [1 mark for correct reactant, 1 mark for correct products]

$C_6H_{12}O_6 + 6O_2 = 6CO_2 + 6H_2O$ [1 mark for correct reactant, 1 mark for correct products]

3. Chemical reactions to build larger molecules from smaller ones; movement/contraction of muscles; to maintain their bodies at a constant temperature. [3 marks]

4. Lipids: made from one molecule of glycerol [1 mark] joined to three fatty acids. [1 mark]

Proteins: glucose and nitrate ions [1 mark] form amino acids, which are used to synthesise proteins. [1 mark]

Anaerobic respiration

1. Normally, we have enough oxygen for aerobic respiration to take place. [1 mark]

Aerobic respiration is preferable as it produces more energy/is more efficient. [1 mark]

2. glucose $\longrightarrow$ ethanol + carbon dioxide. [2 marks]

3. A–B: lactic acid increases slowly in the first 10 seconds but then increases much more rapidly because respiration changes from aerobic to anaerobic because not enough oxygen can reach the muscles.

B–C: lactic acid levels off as he stops accelerating. Aerobic respiration is still occurring but insufficient oxygen reaches the muscles to break down the lactic acid so levels stay high but not increasing.

C–D: rapid increase in lactic acid caused by sprint finish and anaerobic respiration occurring again due to insufficient oxygen reaching muscles.

D–E: lactic acid levels suddenly drop as race has finished, aerobic respiration is happening again and oxygen becomes available to break down lactic acid. [4 marks]

Response to exercise

1. Increased heart rate to pump blood more quickly to deliver the oxygen and glucose needed for respiration in muscles cells so they can contract. Increased breathing rate to increase the amount of oxygen absorbed by the blood and meet demand for extra oxygen in muscle cells. Increased breathing volume to increase the amount of oxygen absorbed in each breath to meet demand for extra oxygen in muscle cells. [3 marks]

2. Blood flowing through the muscles transports the lactic acid back to the liver. In the liver, the lactic acid is converted back to glucose. [2 marks]

3.

Level 3: The response gives a clear and detailed explanation of the relationships between A, B and C, including the relationship between A and C.	5–6
Level 2: The response gives some explanation of the relationships between A, B and C, although explanations may lack some details, such as the relationship between A and C.	3–4
Level 1: The response gives an attempt to explain the relationships between some parts of the graph but the relationships are not fully explained or understood.	1–2
Indicative content: • Area A shows the oxygen deficit. • This is the amount of oxygen that the body is short of to meet the demands of the exercise. • Area B shows the amount of oxygen absorbed by the lungs during exercise. • Area C shows the oxygen debt.	

- This is the extra volume of oxygen needed to pay off the oxygen debt built up during exercise.
- Area A is the same size/area as area C because the oxygen deficit built up during anaerobic respiration has to be paid back in full.

Homeostasis

1. Any **two** from: regulation of internal conditions; to maintain optimum conditions; in response to internal (and external) changes. [2 marks]
2. Any **two** from: carbon dioxide; water; blood glucose levels; body temperature; any other (e.g. urea, salts, named salts). [2 marks]
3. Brain/hypothalamus. [1 mark]
4. To maintain optimal conditions; for enzyme action/cell functions. [2 marks]
5. Nervous system; hormonal system [2 marks]

The nervous system and reflexes

1. Any **three** from: cell body with lots of mitochondria to provide energy; many dendrites to make connections with other neurones; insulating sheath/myelin to prevent electric impulse leaking; long thin nerve fibres to extend over a large area for fast conduction of impulses; synapses/gaps between neurones to prevent low level impulses reaching the CNS/ensuring impulse travels in one direction only. [3 marks]
2. No; all impulses are electrical signal (pulses)/all nerve impulses are similar; impulses from different receptors go to different areas of the brain; when impulses are received in that area, the brain recognises where they have come from. [3 marks]
3. a Any **two** from: the height of the ruler above the hand; the horizontal distance between the thumb and ruler; the same hand to catch each time; same person to catch each time; other valid answer. [2 marks]

   b Any **two** from: caffeine; repetition/learning; distraction/example e.g. loud music; any other valid answer. [2 marks]
4. Any **six** from: automatic response to stimulus; receptor is stimulated; sends impulses to the spinal cord; through sensory neurone; sensory neurone passes impulse to motor neurone; through relay neurone; back to effector (usually muscle); brain cannot stop this reaction. [6 marks]

The brain

1. Neurones. [1 mark]
2. Medulla; cerebral cortex/cerebrum; cerebellum. [3 marks]
3. Studying patients with brain damage; description of famous cases, e.g. Phineas Gage, war wounded, etc.; electrically stimulating parts of the brain; use of animals for electrical stimulations/description of procedure; comparison of diseased and healthy brains; MRI scans/ description of MRI brain scan procedure. [6 marks]
4. a Cerebral cortex/cerebrum. [1 mark]

   b Any **two** from: observing behaviour; testing cognitive function; MRI scans; CAT scans; X-rays. [2 marks]

   c Any **two** from: brain is very complicated; contains billions of neurones; functions are not always localised in one area; one area doesn't just carry out one function. [2 marks]

The eye

1. a

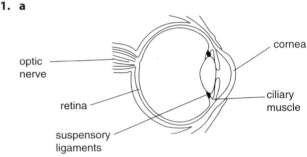

[5 marks]

   b Ciliary muscle; cornea. [2 marks]
2. Any **four** from: (dim) light enters eye; through cornea/pupil/lens; onto retina; converted to electrical impulses; (travel through) optic nerve to brain/ coordination centre; motor impulse to effectors/iris muscles; iris muscles relax; pupil dilates (to allow more light in). [4 marks]
3. Cornea; lens. [2 marks]
4. a The ciliary muscles contract; the suspensory ligaments loosen; the lens is then thicker and refracts light rays strongly. [3 marks]

   b The ciliary muscles relax; the suspensory ligaments are pulled tight; the lens is then pulled thin and only slightly refracts light rays. [3 marks]
5. Diagram showing: a concave/diverging lens; light/rays being diverged before entering eye/pupil; light rays meeting/focusing on retina. [3 marks]
6. Laser surgery to change shape of cornea; replacement of cloudy/defective lenses. [2 marks]

Control of body temperature

1. a Correct calculation of mean: $\frac{814}{22} = 37.0$ [2 marks]

   b Optimum temperature for enzymes; to maintain cell/body functions. [2 marks]

2. Blood cooled by ice; cooled blood detected by thermoregulatory centre in brain; impulses from brain to effectors/named effectors; sweat glands less active; vasoconstriction; shivering/goosebumps/erect hair; therefore, less heat lost by skin. [5 marks]

Hormones and the endocrine system

1. 1 – thyroid (gland); 2 – adrenal (glands); 3 – testis; 4 – pituitary; 5 – pancreas; 6 – ovary. [4 marks]

2. Pituitary; [1 mark] because it secretes several hormones; that regulate secretions of other glands/act on other glands. [2 marks]

3. Any **five** from: composed of glands; that secrete hormones; carried in blood stream; to target organs; to produce an effect; slower but longer-lasting effect (compared to nervous system); named example.

Controlling blood glucose

1. Pancreas; insulin; liver; muscle (**or** muscle; liver); glycogen. [5 marks]

2. Wholemeal bread; produces less glucose (than other breads); glucose is higher than normal for less time (than with other breads); will need less insulin to control glucose levels. [4 marks]

3. Worked example - full answer given in workbook.

Maintaining water and nitrogen balance in the body

1. Exhalation (lungs); sweat (skin); urine (kidneys). [3 marks]

2. a ADH/anti-diuretic hormone. [1 mark]

   b Increases permeability of kidney tubules/increases reabsorption of water. [1 mark]

3. Any **four** from: amino acids deaminated in liver; to form ammonia; ammonia is toxic; immediately converted to urea; dissolved in blood (plasma); transported by blood (to be excreted by kidney). [4 marks]

4. Worked example - full answer given in workbook.

Hormones in human reproduction

1. a Oestrogen, from the ovaries. [1 mark]

   b Testosterone, from the testes. [1 mark]

   c Sperm production. [1 mark]

2. a Pituitary. [1 mark]

   b Ovary. [1 mark]

   c Bloodstream. [1 mark]

3. FSH: causes maturation of an egg/stimulates ovaries to release oestrogen; LH: stimulates the release of an egg. [2 marks]

4. a Rising levels of oestrogen; result in an increased LH level; when LH level peaks, egg release stimulated. [3 marks]

   b Continues to inhibit FSH production and to inhibit LH production; so that no eggs are matured or released; because of danger to a later conceived foetus if two develop in the uterus. [3 marks]

Contraception

1. Birth control pills are 99% effective in preventing pregnancy; the hormones in the pills give protection against some women's diseases; the woman's monthly periods become more regular. [3 marks]

2.

Level 3: A detailed, logical and coherent discussion of positive and negative aspects of the trial.	5–6
Level 2: Some discussion of both positive and negative aspects of the trial.	3–4
Level 1: Some relevant comments about the trial.	1–2
Indicative content: *Positive points:* • large scale trial • showed that pill worked • led to widespread use of pill. *Negative points:* • women not informed it was a trial/experimental drug/about possible side effects • women gave no informed consent • some women affected by side effects • possible deaths due to trial • no pre-clinical testing • no placebo used/no double-blind testing • used high doses/should have started with low doses.	

Using hormones to treat human infertility

1. FSH and LH. [2 marks]

2. a Prolonged use may prevent later ovulation; may cause multiple births. [2 marks]

   b Any **two** from: causes mood swings; named side effects; no protection against STDs; long-term effects. [2 marks]

3. Mother given FSH and LH to stimulate maturation of (several) eggs; eggs collected and fertilised by sperm from father (in the laboratory); fertilised eggs develop into embryo; embryos inserted in mother's uterus. [4 marks]

4. Emotionally/physically stressful; success rates not high; multiple births/risk to babies and mother. [3 marks]

Answers

Negative feedback

1. An increase in blood glucose concentration causes insulin to be secreted; insulin acts to reduce blood glucose concentration; the decrease in blood glucose concentration reduces insulin secretion; blood glucose concentration maintained at an almost steady level. [4 marks]

2. (Produced by adrenal glands) in times of fear/stress/excitement; increases heart rate; boosts delivery of blood/glucose to brain/muscles; prepares body for flight or flight response. [4 marks]

3. a Thyroid (gland). [1 mark]
   b Stimulates basal metabolic rate; role in (mental/physical) growth and development. [2 marks]

Plant hormones

1. a Water/moisture; gravity. [2 marks]
   b Auxin/named auxin. [1 mark]
   c Unequal distribution; causes unequal growth rates. [2 marks]

2. a Shoot **A** unchanged; shoot **B** taller and growing towards the left/light. [2 marks]
   b Shoot **B**: auxin moves down shaded side; increasing growth/cell elongation (on shaded side); shoot **A**: auxin can not move down shaded side/auxin blocked by plastic; no increased growth (on shaded side). [4 marks]
   c Causes plants to grow towards light; more light; more photosynthesis. [3 marks]

3. a No. [1 mark]
   b Worked example - full answer given in workbook.

Uses of plant hormones

1. a Auxin. [1 mark]
   b Roots grow more quickly from cuttings. [1 mark]

2. a Fruit doesn't fall off and become damaged; fruit becomes larger; to speed up ripening/so fruit can be picked at the same time. [3 marks]
   b Gibberellin. [1 mark]

3. a They are firmer; less easily damaged/bruised in transport. [2 marks]
   b Spray/treat with ethene. [1 mark]

4. Worked example - full answer given in workbook.

Section 6: Inheritance, variation and evolution

Sexual and asexual reproduction

1. a Jane. [1 mark]
   b Has horse and zebra features; genes/DNA/chromosomes/genetic information in gametes; zorse receives genes/DNA. [3 marks]

2. Gametes; meiosis; zygote. 2 correct [1 mark] **but** 3 correct [2 marks]

3. a Any **four** from: asexual reproduction: one parent; no gametes/meiosis; no fertilisation/fusion of gametes/mixing of genetic information; no genetic variation in offspring/genetically identical offspring/clones; only mitosis involved. (Or reverse ideas for sexual reproduction.) [4 marks]
   b Dispersal stage/stage in mosquito; needs variation to ensure at least some offspring survive in new/unpredictable environment/host. [2 marks]

4. Any **six** points from: large numbers produced; the plants produced are identical clones; which is an advantage if the environment is constant; giving quicker colonisation; there is a lack of waste, e.g. excess pollen production; so greater chance of survival; and no dependence on insects; water is not required for dispersal of spores/pollen grains; pollen grain is drought-resistant; with adaptations to flowers to ensure transport by wind, insects or other animals; gametes do not need to swim in liquid medium; female gametes retained prior to fertilisation; zygote develops to specialised resistant seeds; seeds resistant to unfavourable conditions, e.g. drought/cold; which normally disperse independently of water, **or** example. [6 marks]

Cell division by meiosis

1. a Meiosis. [1 mark]
   b Mutations in sex cells can be passed to offspring (while mutations in body cells cannot). [1 mark]

2. a Pollen **and** egg cell/ovum. [1 mark]
   b Sperm **and** egg cell/ovum. [1 mark]
   c 46; 23; 46. [3 marks]

3. Any **six** comparisons from:

meiosis	mitosis
sexual	asexual
gametes	growth
ovary or testes or gonads	all other (body) cells
half number of chromosomes or haploid	same number of chromosomes
23 chromosomes	46 chromosomes
reassortment	no reassortment
variation possible or not identical	no variation or identical
4 cells produced	2 cells produced
2 divisions	1 division

[6 marks]

DNA, genes and the genome

1. 23; 40; 7; 210. [2 marks]

2. The entire genetic material of an organism. [1 mark]

126

3. **a** Some people might not want to know; they may be discriminated against/valid example e.g. may not be able to get insurance. [2 marks]

   b Understand/treat inherited disorders; trace human migration patterns. [2 marks]

4. Chemical reactions determine type/functions of cells; controlled by enzymes; enzymes are proteins; sequence of bases in DNA; determines which amino acids are used/order they are joined; to form protein (therefore genes determine cell structure). [6 marks]

Structure of DNA

1. Common sugar/5-carbon sugar/deoxyribose molecule; joined to phosphate; and (organic) base/four types of bases/ACTG. [3 marks]

2. **a** The phosphate group of one nucleotide joins with the sugar of the next nucleotide. [1 mark]

   b Three. [1 mark]

   c Code for different amino acid; leading to a change in protein. [2 marks]

3. **a** Can switch genes on and off/regulate gene expression. [1 mark]

   b It will alter gene expression/change phenotype/ harmful or beneficial mutation. [1 mark]

4. **a** They must be paired together; T to A and C to G. [2 marks]

   b Any **four** from: genes are sections of DNA; (genes) code for the amino acid sequence of a protein; consist of bases/four different bases/A, T, G and C; order of bases codes for order of amino acids; triplet code. [4 marks]

Protein synthesis and mutations

1. **a** TACATGGAT All A and T paired up; all C and G paired up. [2 marks]

   b Three/3. [1 mark]

2. At the ribosomes. [1 mark]

3. Any **six** from: DNA (in nucleus) unzips; sequence of bases in a gene acts as template to produce messenger molecule/mRNA; mRNA leaves nucleus attaches to ribosome; carrier molecule/tRNA brings in an amino acid; complementary base pairing attaches tRNA to ribosome; second amino acid arrives at ribosome and amino acids attach; tRNA leaves ribosome; idea that protein chain starts to grow as process is repeated as messenger molecule moves across ribosome. [6 marks]

4. **a** Any **two** from: specific shape; to bind substrate/ form enzyme-substrate complex; for reaction to take place; reference to lock and key mechanism/ hypothesis. [2 marks]

   b Worked example - full answer given in workbook.

Inherited characteristics

1. **a** Different/alternative version of a gene. [1 mark]

   b Spotted leopard: BB or Bb; black panther: bb. [3 marks]

   c

		Spotted leopard	
		B	b
Black panther	b	Bb spotted	bb black
	b	Bb spotted	bb black

   Probability = 50%/0.5/½/1 in 2

   Parental alleles correct; offspring genotypes correct; offspring phenotypes correct; correct probability. [4 marks]

2.

Level 3: A clear explanation, including correct probabilities, supported by an appropriate Punnett square diagram.	5–6
Level 2: A clear explanation supported by an appropriate Punnett square diagram.	3–4
Level 1: A partial explanation including an attempt at an appropriate Punnett square diagram.	1–2

   Indicative content:

		Long wing (first generation)	
		L	l
Long wing (first generation)	L	LL long wing	Ll long wing
	l	Ll long wing	ll short wing

   - Identification of long wing as dominant condition and short wing as recessive condition.
   - Identification of long wing first generation as heterozygous.
   - Use of suitable symbols: uppercase for long wing and lowercase for short wing.
   - Correct Punnett square diagram showing cross between heterozygous long wing flies.
   - Correct identification of the genotypes and phenotypes of the offspring of this cross.
   - Long wing and short wing offspring in the ratio 3 : 1.

Inherited disorders

1. Sam = Dd; mother = Dd; father = dd; sister = dd. [4 marks]

2. Any/all children will be Ff; (so) will not have cystic fibrosis; (but) will be carriers; (there is a) possibility of grandchildren having cystic fibrosis if a child's partner has cystic fibrosis/is a carrier. [4 marks]

Sex chromosomes

1. **a** XX [1 mark]

    b XY [1 mark]

2. Wrong; [no mark] any **two** from: boy is XY, girl is XX; equal numbers of XX and XY; 50%/equal chance of having boy or girl. [2 marks]

3. A gene on the Y-chromosome determines the sex of the embryo; the sex-determining gene triggers the development of testes; each body cell in a human female contains XX sex chromosomes. 2 correct [1 mark] **but** 3 correct [2 marks]

4. To produce a male child a Y chromosome is needed; Y chromosomes are only found in (50% of) sperm/all eggs carry X chromosomes. [2 marks]

Variation and mutations

1. Differences between members of the same species/ organisms of the same kind. [1 mark]

2. Environmental factor – any **one** from: light; water; rain; temperature; minerals; space. [1 mark]

    Other factor: genes/inheritance. [1 mark]

3. Use of the antibiotic kills non-resistant bacteria; the antibiotic-resistant bacterium survives and multiplies; antibiotic-resistant bacteria spread/become more common in the population. [3 marks]

4. **a** 1 only. [1 mark]

    b 12. [1 mark]

    c Any **one** from: fossil records do not always form; soft tissues/skin/colour are not preserved; many fossils are yet to be found. [1 mark]

Theory of evolution

1. A change in the (inherited) characteristics of a population; over a (long) period of time. [2 marks]

2. **a** Organisms produce more offspring than can survive, leading to competition within a population/struggle for survival; within a population there is (wide) variation; individuals with characteristics most suited to the environment are more likely to survive/survival of the fittest; these individuals are more likely to breed and pass on their characteristics/alleles to offspring; over time the population as a whole becomes better adapted/change in allele frequency. [5 marks]

    b Idea that changes that occur during an organism's lifetime can be passed on to offspring/inheritance of acquired characteristics. [1 mark]

    c (Most) changes occurring to an organism during its lifetime are not inherited; because the genes are not affected. [2 marks]

3. In woodland dark snails are better camouflaged; less likely to be eaten; in grassland pale snails are better camouflaged; less likely to be eaten. Allow reverse arguments. [4 marks]

4. Worked example - full answer given in workbook.

Speciation

1. **a** Galapagos finches descended from species on the mainland; different conditions/foods on each island; on each island finches evolved by natural selection/become better adapted to conditions on each island; eventually finches on the separate islands became so different they could no longer interbreed (successfully) (and so had become separate species). [4 marks]

    b The theory challenged the idea that God created all species; mechanism of inheritance/variation/genes not known (at the time); more evidence discovered over time. [3 marks]

2. When the forest split into smaller forests, populations (of the same original species) were separated and could no longer interbreed; different mutations occurred in each population/different environmental conditions in each forest; natural selection selects for different characteristics in different forests/populations evolve independently in different forests and develop different phenotypes; when forests join together populations (of same original species) can now mix; however, they have become so different they can no longer interbreed successfully/to produce fertile offspring so separate species have formed/speciation has occurred; this process is repeated many times. [6 marks]

The understanding of genetics

1. **a** All the offspring are tall. [1 mark]

    b Yes, results are similar/no outliers; consistent 3:1 ratio. [2 marks]

2. **a** Any **two** from: purple flowers; yellow seeds; green seed pods. [2 marks]

    b Use of 3:1 ratio in answer; 450. [2 marks]

    c Cross short white-flowered plants together; because short and white flowers are both recessive characteristics. [2 marks]

3. Any **three** from: other scientists can build upon their result; so can develop ideas quicker; other scientists can repeat/test the work; different teams have different skills/resources/ideas/approaches; so that a broad range of evidence can be put together to develop the idea. [3 marks]

Fossil evidence

1. **a** Evolution. [1 mark]

   b Any **two** from: parts of organism that have not decayed; parts of organism replaced by minerals; preserved traces/footprints/burrows/other named traces. [2 marks]

   c Gaps in fossil record so lack of evidence of evolution into modern penguins; it could be a close relative but not a direct ancestor. [2 marks]

2. Any **three** from: soft-bodied organisms/soft tissues do not easily form fossils; many species did not live/die in environments suitable for forming fossils; many fossils have not yet been found; many fossils have been destroyed, e.g. by erosion/the rock cycle. [3 marks]

Other evidence for evolution

1. Increased use of antibiotics; random changes in the genes of microorganisms. [2 marks]

2. Worked example - full answer given in workbook.

3. **a**

Level 3: Detailed explanation covering mutation **and** antibiotic resistance **and** natural selection.	5–6
Level 2: Partial explanation covering mutation **or** antibiotic resistance **and** natural selection.	3–4
Level 1: Partial explanation covering mutation **or** antibiotic resistance **or** natural selection.	1–2
No relevant content	0
Indicative content: • Bacteria develop (random) mutations. • Introduces new strains into population. • Some strains less affected by particular antibiotics/antibiotic resistant. • Antibiotics kill non-resistant strains/resistant strains are selected for. • Reduced competition. • Resistant strains survive and reproduce. • New strain becomes more common in the population.	

   b Any **two** from: do not overuse antibiotics/do not use when inappropriate, e.g. for non-serious/viral infections; patients to complete courses of antibiotics; restrict agricultural use of antibiotics. [2 marks]

Extinction

1. **a** That there are no remaining/living individuals left. [1 mark]

   b Any **four** from: changes to the environment; new predators; new diseases; new (more successful) competitors; speciation; catastrophic event/described event. [4 marks]

   c Human predation; melting of ice (destroying mammoth habitat). [2 marks]

Selective breeding

1. Any valid characteristic and suitable explanation, e.g. a gentle nature – so will not cause harm; or obedient – so can follow instructions. [2 marks]

2. **a** Apples that are large **and** sweet. [1 mark]

   b No, he cannot be sure (no mark); he may get other combinations of characteristics/small and not so tasty. [1 mark]

   c So the flowers were not pollinated by pollen from other trees/were only pollinated by pollen from apple tree B. [1 mark]

   d Plant the seeds so they grow into trees; select the trees that produce the best fruit; cross-breed these trees together; repeat the whole process many times. [4 marks]

3. **a** Tomatoes will be produced more quickly/can sell them before competitors/can get more crops in a season. [1 mark]

   b Less (genetic) variation; fewer different alleles/smaller gene pool; because selection removes some alleles from population. [3 marks]

Genetic engineering

1. Genes from one organism are transferred to a different organism. [1 mark]

2. **a** To insert gene into (bacterial) required cells. [1 mark]

   b Plasmid [1 mark]

   c An enzyme. [1 mark]

   d A different enzyme. [1 mark]

3. **a** Virus is rendered harmless. [1 mark]

   b Any **three** from: reversion of virus to disease-causing form; bacteria/virus could be toxic to humans or insects; virus/bacterium could transfer to another species; accept a reference to ethical reasons, e.g. why change colour of leaves when there is no nutritional value? [3 marks]

4. Any **three** from: bacteria reproduce quickly/can be grown in large numbers; to produce large amounts of insulin; less likely to cause reaction; overcomes concerns from vegetarians/religious people. [3 marks]

Cloning

1. **a** 1, 4, 5, 2, 3. Three/four in correct sequence = [1 mark] **but** all correct = [2 marks]

   b Can produce more offspring; because use (several/many) surrogates. [2 marks]

2. **a** C/tissue culture. [1 mark]

   b Plants grown from seeds may not be disease-resistant; plants grown from seeds will not be clones/genetically identical to original plant; tissue culture produces more clones than cuttings can/can only take a (relatively) small number of cuttings from one plant. [3 marks]

3. **a** Remove nucleus from (elephant) egg cell; nucleus from (mammoth) skin cell inserted into (elephant) egg cell; electric shock stimulates egg cell to divide to form an embryo; embryo inserted into womb/uterus of female elephant. [4 marks]

   b Only has genes/DNA from mammoth. [1 mark]

Classification

1. **a** Putting things into groups. [1 mark]

   b (Carl) Linnaeus. [1 mark]

   c Name made of two parts; genus and species. [2 marks]

   d Group of organisms with similar features/characteristics; capable of reproducing fertile/viable offspring. [2 marks]

2. **a** Any **two** from: animals; plants; fungi. [2 marks]

   b Archaea; bacteria. [2 marks]

   c Chemical analysis/RNA sequencing. [1 mark]

Section 7: Ecology

Habitats and ecosystems

1. A habitat is the place where a particular organism lives; all the populations of different organisms that each have their own habitat make up a community; the community is made up from different living organisms that interact with non-living organisms, and together these form an ecosystem. [3 marks]

2. **a** Correct type of graph drawn (line graph); day on x-axis and number of duckweed plants on y-axis, both labelled correctly; appropriate scales chosen; accurate plotting of data. [4 marks]

   b The number of duckweed plants increases rapidly/exponentially between day 0 and day 25; the population then stops increasing and becomes stable; likely to be due to availability of space, light or carbon dioxide limiting further growth. [3 marks]

Food in the ecosystem

1. **a** Primary consumer. [1 mark]

   b The direction of the transfer of energy from one organism to another. [1 mark]

2. A food web is more accurate as it shows the network of interactions between organisms.

   A food web shows all or many of an organism's predators and prey whereas the food chain only shows one. [2 marks]

3. Because so much biomass/energy is lost from the food chain at each stage; through uneaten material, waste products (urea and CO_2), faeces, energy for movement and keeping warm. [2 mark]

4. The Sun/light source provides the energy for photosynthesis; photosynthesis makes glucose and oxygen; the glucose serves as food for animals; the oxygen is used for respiration by living organisms living in the ecosystem. [4 marks]

Abiotic and biotic factors

1. Any **four** from: amount of sunlight/shade; temperature; wind; moisture; frost or snow. [4 marks]

2. Any **three** from: preying on native species; out-competing native species for food or other resources; causing or carrying disease; preventing native species from reproducing or killing their young. [3 marks]

3. Because both predator and prey mites are responding to changes in the previous generation of the other population; because neither population can respond instantly to changes in the other population; because reproductive cycles in either population cause a delay in response to the other population. [3 marks]

4. **a** It is more resistant to drying out than *Balanus*; the upper shore is left uncovered by the tides for long periods each day; and only the *Chthamalus* can tolerate long periods of exposure to air and the drying action of the sun. [3 marks]

   b The *Balanus* are faster growing and out-compete the *Chthamalus* for space; the *Balanus* out-compete the *Chthamalus* for food. [2 marks]

Adapting for survival

1. Structural adaptations are physical features of an organism that help it survive; behavioural adaptations are the things organisms do to survive; functional adaptations are any changes to the way an organisms works/functions that helps it become more favourable to a certain environment. [3 marks]

2. **a** Organisms that live in very extreme environments, such as high temperature, pressure or salt concentration. [1 mark]

   b Fewer or no predators. [1 mark]

3. Flexible stems – allow plant to bend with flow of water without getting damaged; aerial roots – allow plant to get oxygen for root respiration and needed for uptake of minerals. [2 marks]

4. Worked example - full answer given in workbook.

Measuring population size and species distribution

1. **a** Quadrat 4, test 2 circled. [1 mark]

   b Mean values: quadrat 1 = 24.0, 2 = 35.4, 3 = 43.5 and 5 = 69.6; [1 mark] quadrat 4 = 56.6 (anomalous result should be ignored); [1 mark] all numbers rounded to one decimal place. [1 mark]

   c As the soil moisture increased, the percentage cover of creeping buttercups increased.

   As the soil moisture decreased, the percentage cover of bulbous buttercups increased. [2 marks]

   d No; his results do support his hypothesis, but do not prove anything; one investigation is not enough to prove a hypothesis. [2 marks]

   e Collect data every metre along the transect line; repeat procedure in another area of meadow where ground goes from wet to dry; use larger quadrat; use quadrat divided into 100 squares and estimate exact percentage cover. [3 marks]

Cycling materials

1. Water evaporates from the land and sea using energy from the Sun; water also evaporates from plants during transpiration and becomes water vapour; the water vapour rises up and cools due to lower pressure, and condenses forming clouds; eventually the water falls as precipitation (or snow/hail) onto the land before draining back into the sea. [4 marks]

2. Every living organism on Earth depends on water to survive; without water and the water cycle to circulate water, all living organisms would die very quickly; it is needed for chemical reactions in living organisms such as respiration and photosynthesis. [3 marks]

3. Any **two** from: different seasons would mean different amounts of photosynthesis occurring and therefore different rates of carbon dioxide removal; amount of carbon dioxide stored in oceans might change according to air temperature; amount of combustion taking place could also affect the amount of carbon dioxide being released. [2 marks]

4. No microorganisms would mean no decomposition; carbon would not be returned to the soil or released back into the atmosphere from respiration of microorganisms/carbon would not be cycled. [2 marks]

5.

Level 3: The response gives a clear and detailed account of the carbon cycle.	5–6
Level 2: The response gives a basic account of the carbon cycle.	3–4
Level 1: The response gives very basic information about the carbon cycle, but parts of the cycle might be incomplete.	1–2
Indicative content:	

- Carbon enters the atmosphere as carbon dioxide from respiration and combustion.
- Carbon dioxide is absorbed by producers to make carbohydrates in photosynthesis.
- Animals feed on the plant, passing the carbon compounds along the food chain.
- Some carbon consumed is exhaled as carbon dioxide formed during respiration.
- Animals produce waste, which is broken down by microorganisms.
- The animals and plants eventually die.
- The dead organisms are decomposed by microorganisms.
- Microorganisms respire, releasing carbon dioxide back into the atmosphere.
- Carbon in the bodies of dead plants/animals is returned to the atmosphere as carbon dioxide.
- In some conditions the plant and animal material may become fossil fuels.
- Fossil fuels release carbon back into the atmosphere in combustion.
- Combustion of carbon- based products such as trees also releases carbon dioxide into the atmosphere.

Decomposition

1. Sunny area – increases temperature and rate of enzyme-controlled reactions in microorganisms involved in decay and therefore speeds up rate of decay; keeping it moist – microorganisms involved in

decay need moisture to carry out biological processes so keeping the lid on ensures decomposition can take place; mixing the contents – increases the oxygen which microorganisms involved in decay need to respire. The more oxygen, the faster the rate of decay. [3 marks]

2. Worked example - full answer given in workbook.

3. Without decay the vital nutrients would remain locked inside dead organic materials. [1 mark]

4. To provide insulation/less temperature variation/maintain temperature; less chance of microbes being killed/enzymes denatured **or** keep at optimum temperature **or** maintain high gas production. [2 marks]

Changing the environment

1. Any **three** from: less food available for organism that rely on trees for food; less shelter/habitats available for organisms; less transpiration can affect microclimate/water cycle by reducing rainfall so less water might be available to organisms; soil becomes less fertile due to disruption of nutrient cycles so fewer plant species able to grow; increased soil erosion and landslides can waste away fertile soil/remove habitats. [3 marks]

2. Increase in salinity: species not tolerant to change in salt would decrease/proliferation of species adapted to survive in more saline environments; decrease in oxygen: decrease in marine life due to less oxygen concentration; decrease in decomposition if insufficient oxygen for decomposers and therefore there is an accumulation of remains of dead plants and animals; increase in sea temperature: species not tolerant to higher temperatures would die/increase in species better adapted to warmer sea temperature. [3 marks]

3. Global warming would increase the air and sea temperatures; which could cause a significant increase in the rate of photosynthesis in the phytoplankton; because the data shows that an increase from –1 °C to 3 °C increases the rate of photosynthesis from 100 au to 143 au; which is a large increase for quite a small change in temperature. [4 marks]

Effects of human activities

1. The variety of different species of organisms on Earth or within an ecosystem. [1 mark]

2. Biodiversity makes an ecosystem more stable; by reducing the dependence of one species on another for food, territory or mates; some species are interdependent and one species may support the survival of another, e.g. by helping maintain the right physical conditions. [3 marks]

3. Increased levels of phosphates and nitration could cause coastal eutrophication; sewage containing dangerous microorganisms could be carried onshore by currents or winds causing pollution, disease and an unpleasant odour. [2 marks]

4. Advantages: waste products are disposed of in managed/monitored landfill sites and not dumped; waste can be processed and suitable materials recycled. [2 marks]

   Disadvantages: any **two** from: they take up space and may destroy habitats; they are unsightly/cause visual pollution; the areas surrounding the landfills become heavily polluted and toxic chemicals can leach out into groundwater system; reduce biodiversity of an area. [2 marks]

   Final mark for judgement/evaluation of points raised. [1 mark]

Global warming

1. Any **two** from: burning fossil fuels; more rice crops; increase in cattle farming; deforestation; destruction of peatlands; more petrol cars being used; any other appropriate answer. [2 marks]

2. Rice crops; cattle farming. [2 marks]

3. **a** The graph shows that the temperature has fluctuated but overall there is an increase in global temperature. The graph clearly shows an overall trend of increasing global temperatures of about 0.7 °C between 1880 and 2000. [2 marks]

   b The graph provides strong evidence that global warming is happening as there is a clear trend of temperatures rising. However, the graph also shows some periods when the global temperature decreases/evidence from graph given. The warming trend could also still be part of a longer term cycle of natural ups and downs due to natural changes such as sunspots, so global temperatures may decrease again in the future. Overall, the evidence for global warming shown by the graph is quite strong/accept other judgement based on evidence presented. [4 marks]

Maintaining biodiversity

1. Hedgerows provide a habitat for a wide range of plants and animals, e.g. wildflowers, insects and birds; field margins allowing wildlife to move freely between habitats to find food, shelter, mates; wildflowers in the hedgerow and margin are important sources of nectar and pollen for insect-pollinators so field margins promote the pollination of plant species dependent on insect-pollinators. [3 marks]

2. The natural habitat for some African elephants is turned into cropland by locals due to an increasing population and demand to produce more food from crops to feed local populations and make money. Locals may see this as necessary to survive whereas conservation programmes want to stop destruction of the elephant habitat and want to make sure the elephants' habitat and migratory routes are not broken up. Some locals poach elephants for their tusks, which can sell for a lot of money. The meat from elephants

can also sell for a good price. Locals, that might have no other source of income, can earn a lot of money from poaching. Conservation groups put effort into stopping poaching and educating people about the cruelty involved and the need to protect the elephant species. However, locals might not understand the reasons to protect a species and the need to earn money and bring food to the table is more important to them than protecting the elephant. [4 marks]

3.

Level 3: The response gives a clear and detailed explanation of a wide range of benefits of protecting the rainforest.	5–6
Level 2: The response gives an explanation of most of the benefits of protecting the rainforest.	3–4
Level 1: The response gives a basic explanation of some of the benefits of protecting the rainforest.	1–2
Indicative content: • Maintain the very high biodiversity of the rainforest and prevent endangering/extinction of species. • Maintaining biodiversity of the rainforest increases the stability of the ecosystem due to interdependence of species. • Maintaining biodiversity also would mean that we could continue to use plant products for medicines and continue to discover new ones. • Less carbon dioxide released into the atmosphere compared to if areas were deforested through slash and burn. • Less carbon dioxide released by respiration of microorganisms feeding on dead plant material left behind from deforestation. • More carbon dioxide removed from the atmosphere in photosynthesis so less contribution to greenhouse effect and global warming. • Increased greenhouse gas methane in the atmosphere if land is used for rice growing or cattle ranches. • Water cycle and microclimate are maintained so environment is more stable.	

Biomass in an ecosystem

1. Trophic level 1: contains producers such as plants and algae; which make their own food by photosynthesis from energy from the Sun. [2 marks]

Trophic level 2: contains primary consumers; which are herbivores that eat the plants and algae. [2 marks]

2. 1 mark for drawing pyramid in correct order with trophic level 1 at bottom and labelled correctly; 1 mark for drawing levels to scale.

3. Pyramids of numbers and pyramids of biomass only show the organisms present at a particular time, seasonal differences are not apparent; **or** zooplankton eat the phytoplankton so quickly that the phytoplankton never gets the chance to attain a large biomass. [1 mark]

4. a Energy transfer between trophic levels 2 and 3 = $760 \div 7900 \times 100 = 9.6\%$. [1 mark]

Energy transfer between trophic levels 3 and 4 = $63 \div 760 \times 100 = 8.3\%$. [1 mark]

b Not all of the organisms in one trophic level are eaten by the next trophic level. Some parts may be inedible; some organisms may die before they get eaten so their remains decompose and their energy gets passed to the decomposers; not all the food that is eaten is digested and not all digested food is absorbed, some is egested as faeces; most of the energy is lost through excretion (CO_2 and urea) and cell respiration to provide energy for movement and keeping warm rather than to make biomass [4 marks]

Food security

1. Having enough food to feed a population. [1 mark]

2. Any **four** from: increasing birth rate; changed diets in developed countries; new pests and pathogens affecting farming yields; environmental changes that affect farming, e.g. drop in rainfall; conflict/war. [4 marks]

3. Fishing quotas limit numbers and size of fish that can be caught in an area, preventing particular species being overfished; this helps to maintain fish stocks at a sustainable level to allow breeding to occur to prevent decline and possible extinction of species. [2 marks]

4. It could increase global food security; because far less energy is needed to produce food products from plants; examples of evidence from table given to support points. [3 marks]

Role of biotechnology

1. Mycoprotein/Quorn®. [1 mark]

2. Advantages: they increase the crop yield as resistant to pests or herbicides; they can be engineered to have improved nutritional value or contain nutrients lacking in a population.

Disadvantages: they won't survive on poor soils and can't solve the food shortage in all areas; loss of biodiversity as fewer weed species survive as a food and shelter source for animals. [4 marks]

3.

Level 3: The response considers most of the advantages and disadvantages, and gives an opinion about use of biotechnology to produce insulin based on the points raised.	5–6	
Level 2: The response considers some of the advantages and disadvantages, and gives an opinion about use of biotechnology to produce insulin based on the points raised.	3–4	
Level 1: The response considers at least one advantage and one disadvantage. There may be an attempt at a concluding opinion but it may not be supported by the points raised.	1–2	
Indicative content: Advantages: • Human insulin can be produced in large quantities. • The process is not time-consuming and is inexpensive compared to previous methods.		

- The insulin is indistinguishable from human insulin produced in the pancreas, and therefore is less likely to cause allergic reactions in people with diabetes.
- The insulin is produced in a very controlled and sterile environment and then purified to ensure it is pure and of the required faster.
- The insulin is absorbed more rapidly than animal-derived insulin and acts faster.

Disadvantages:
- The production costs are still high.
- Some people think that genetic engineering is not ethical.